OCEAN CHEMISTRY AND DEEP-SEA SEDIMENTS

PREPARED BY AN OPEN UNIVERSITY COURSE TEAM

PERGAMON PRESS
OXFORD · NEW YORK · BEIJING · FRANKFURT · SÃO PAULO · SYDNEY
TOKYO · TORONTO

in association with

THE OPEN UNIVERSITY
WALTON HALL, MILTON KEYNES, MK7 6AA, ENGLAND

U.K.	Pergamon Press plc, Headington Hill Hall, Oxford OX3 0BW, England
U.S.A.	Pergamon Press Inc., Maxwell House, Fairview Park, Elmsford, New York 10523, U.S.A.
PEOPLE'S REPUBLIC OF CHINA	Pergamon Press, Room 4037, Qianmen Hotel, Beijing, People's Republic of China
FEDERAL REPUBLIC OF GERMANY	Pergamon Press GmbH, Hammerweg 6, D-6242 Kronberg, Federal Republic of Germany
BRAZIL	Pergamon Editora Ltda, Rua Eça de Queiros, 346, CEP 04011, Paraiso, São Paulo, Brazil
AUSTRALIA	Pergamon Press Australia Pty Ltd., P.O. Box 544, Potts Point, N.S.W. 2011, Australia
JAPAN	Pergamon Press, 5th Floor, Matsuoka Central Building, 1-7-1 Nishishinjuku, Shinjuku-ku, Tokyo 160, Japan
CANADA	Pergamon Press Canada Ltd., Suite No. 271, 253 College Street, Toronto, Ontario, Canada M5T 1R5

First edition 1989

Library of Congress Cataloging in Publication Data
Ocean chemistry and deep-sea sediments / prepared by an Open University course team. 1st ed.
p. cm.
Bibliography: p.
Includes index.
1. Chemical oceanography. 2. Marine sediments. I. Open University.
GC111.2.034 1989 551.46′01—dc20 89–8456

British Library Cataloguing in Publication Data
Ocean chemistry and deep-sea sediments
1. Oceans. Chemical composition & Chemical properties
I. Open University
551.46′01

ISBN 0-08-036374-1 Hardcover
ISBN 0-08-036373-3 Flexicover

Jointly published by the Open University, Walton Hall, Milton Keynes, MK7 6AA and Pergamon Press plc, Headington Hill Hall, Oxford OX3 0BW.

Designed by the Graphic Design Group of The Open University.

Printed in Great Britain by BPCC Wheatons Ltd., Exeter, England.

CONTENTS

ABOUT THIS VOLUME

This is one of a Series of Volumes on Oceanography. It is designed so that it can be read on its own, like any other textbook, or studied as part of S330 *Oceanography*, a third level course for Open University students. The science of oceanography as a whole is multidisciplinary. However, different aspects fall naturally within the scope of one or other of the major 'traditional' disciplines. Thus, you will get the most out of this Volume if you have some previous experience of studying geology and a certain amount of chemistry. Other Volumes in this Series lie variously within the fields of geology, biology, physics or chemistry.

Chapter 1 begins by explaining why the study of deep-sea sediments is of contemporary interest, and then summarizes the distribution and nature of deep-sea sediments. These include calcareous and siliceous oozes formed from the remains of planktonic organisms, as well as products of terrestrial weathering brought to the sea by rivers, wind and ice.

Chapter 2 reviews the steady-state ocean concept and explains how different constituents can be classified according to their involvement in the biological particle cycle (the formation, destruction and regeneration of organic matter), which is also a major control on residence times in the oceans. A simple two-box model of ocean cycling is introduced and the role of dissolved gases is described, with special reference to oxygen.

Chapter 3 discusses how water chemistry controls the extent to which the remains of calcareous and siliceous organisms dissolve when they sink to the sea-bed and form oozes. The preservation of calcareous and siliceous remains enables aspects of the history of oceans to be deduced.

Chapter 4 describes how the sediments that come from continental areas are initially deposited mainly on continental shelves, and are then transported to the deep sea by turbidity currents and other gravity flows.

Chapter 5 provides a review of processes occurring at and near the sea-bed. The uppermost layers of sediment are disturbed and disrupted by the activities of marine organisms (deep-sea benthos) and by the effects of strong bottom currents which can resuspend and redistribute sedimentary material. In addition, there are important chemical exchanges between seawater and sediment, not only at the interface itself, but within the main body of the sediment pile.

Finally, you will find questions designed to help you to develop arguments and/or test your own understanding as you read, with answers provided at the back of this Volume. Important technical terms are printed in **bold** type where they are first introduced or defined.

CHAPTER 1 INTRODUCTION

'He had long ago decided, since he was a serious scholar, that the caves of ocean bear no gems, but only soggy glub and great gobs of mucky gump.'

James Thurber

Most substances that enter the oceans ultimately end up in the sediments. On the way, they participate in a variety of complex biological and chemical cycles and interactions which involve some substances more than others. Interactions continue after deposition: sediments do not lie passively on the sea-bed until they are buried. Deep-sea animals disturb the sediments as they forage for food, and some sediment may experience erosion and resuspension by bottom currents before being redeposited and finally buried. Chemical reactions may occur between the mineral grains and the overlying seawater, and these reactions can continue after burial, when seawater becomes trapped among the grains.

Nowadays, deep-sea sediments are a focus of much research effort, because of the growing need to quantify the various fluxes contributing to the global carbon cycle. A principal aim is to find out just what happens to the anthropogenic carbon dioxide and other gases contributing to the **greenhouse effect**, which may already have begun to lead to global atmospheric warming. If the present trend continues, major climatic changes and rising sea-levels will result. To predict the extent of such changes successfully, it is essential to know more about rates of increase of greenhouse gases in the atmosphere. This in turn requires improved knowledge of, for example, seasonal, inter-annual and regional fluctuations in ocean surface productivity, the removal of organic matter to the sea-bed, exchanges across the air–sea interface, and vertical and horizontal water movements. International and national programmes initiated to investigate these and related problems include the Joint Global Ocean Flux Study (JGOFS), the Biogeochemical Ocean Flux Study (BOFS), and the somewhat broader **World Ocean Circulation Experiment (WOCE)**.

Nor should the seawater itself be forgotten. The dissolved constituents of seawater include metals such as copper, lead, zinc, tin, manganese, cadmium, mercury, nickel and silver. Geochemical cycles of these and other heavy metals are being grossly perturbed by human activity. Inputs have increased substantially since the Industrial Revolution, although only in the case of lead has the average concentration in open ocean waters actually increased, as a result of atmospheric fall-out. In addition, there are many new substances in the marine environment that were not there even a century ago. These include pesticides and other organic chemicals, as well as transuranic and other 'man-made' nuclides from nuclear weapons testing and low-level waste discharges. Global flux studies will contribute greatly to our understanding of how the many constituents of seawater—both dissolved and particulate, both natural and artificial—move through the various marine chemical cycles. But we begin by looking at the sediments.

1.1 THE DISTRIBUTION OF DEEP-SEA SEDIMENTS

Figure 1.1 HMS *Challenger*, 1872. She was a steam-assisted wooden corvette of 2306 tonnes.

When HMS *Challenger* (Figure 1.1) returned to England on 24 May 1876 laden with specimens, records and measurements after an epic three years and nine months voyage of oceanic exploration, the era of systematic oceanography had begun. The member of the scientific party who did most to ensure world recognition of the *Challenger*'s scientific achievements was John Murray, a Canadian-born Scot, who owed his place on the ship to chance, when a member of the original team was obliged to drop out at short notice. Murray's account of the samples collected from the floor of the oceans (Figure 1.2) provided the starting point for all subsequent investigations into deep-sea sediments.

Figure 1.2 Drawings of planktonic remains in sediments collected from the North Atlantic during *Challenger's* voyage.

(a) Drawings of Radiolaria of the genus *Hexastylus*.

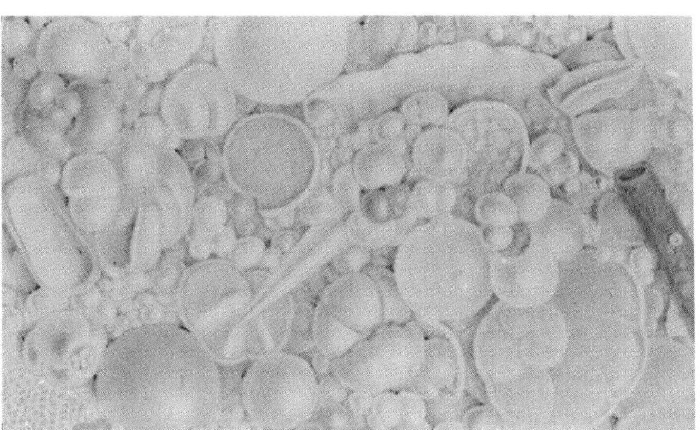

(b) Washed foraminiferal ooze, mainly *Globigerina* spp. (hence the term *Globigerina* ooze). Other shells include a pteropod (centre, see (c)) and an ostracod (crustacean bivalve, extreme left centre). The brown tube (right centre) was probably secreted by an agglutinating (arenaceous) foraminiferan species. Width of field about 3.5mm. Sample from 1900 fathoms (*c.* 3500m).

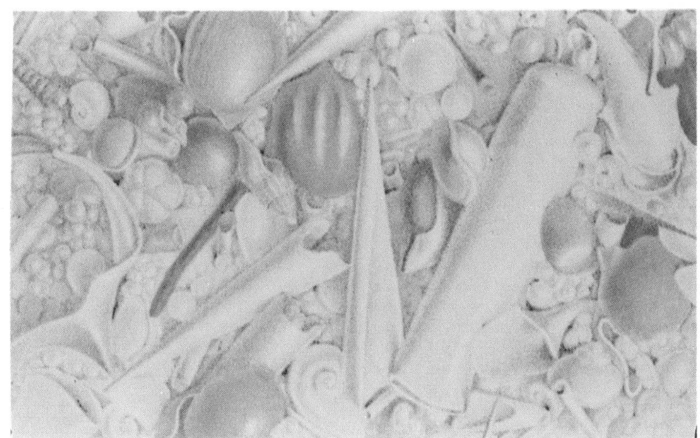

(c) Washed pteropod ooze. The large cone-shaped shell and some of the larger coiled shells are from pteropods (planktonic gastropods). Foraminiferan remains (see (b)) make up the rest. Width of field 18mm. Sample from 450 fathoms (*c.* 800m).

Figure 1.3 Some deep-sea sediment cores,

(a) *Top:* Chocolate-coloured terrigenous 'red clay', typically structureless and with few organic remains other than occasional fish teeth.

Bottom: Diatom ooze from the Antarctic Ocean. The high concentration of diatoms gives the sediment a 'fluffy' aspect—when dry, it looks like glass wool.

(b) *Top:* Mixed radiolarian and calcareous ooze, typical of tropical oceans, especially the tropical Pacific. The mottling is due to the burrowing activities of benthic organisms (bioturbation).

Bottom: Calcareous ooze, formed of coccoliths and foraminiferal remains, between them making up about 90% of the whole.

At first sight, deep-sea sediments are little more than soft muds of various hues, from white to grey to reddish-brown (Figure 1.3). Two main kinds of deep-sea sediments can be recognized.

Terrigenous sediments are formed by weathering and erosion of land areas, and are transported to the oceans by rivers, glaciers and wind. They comprise gravels, sands, silts and clays. Biogenic (or biogenous) sediments are made up of the microscopic remains of those predominantly **planktonic** marine organisms that secrete skeletons (or tests) of calcium carbonate or silica. Figure 1.4 and Table 1.1 summarize the distribution of deep-sea sediments, and Figure 1.5 is a map of the major physiographic features of the ocean basins.

Table 1.1 Percentage of deep ocean floor covered by pelagic sediments.

Sediments	Atlantic	Pacific	Indian	World
Calcareous ooze	65.1	36.2	54.3	47.1
Pteropod ooze	2.4	0.1	—	0.6
Diatom ooze	6.7	10.1	19.9	11.6
Radiolarian ooze	—	4.6	0.5	2.6
Pelagic clays	25.8	49.0	25.3	38.1
Relative size of ocean (% of total)	23.0	53.4	23.6	100.0

QUESTION 1.1 Examine Figures 1.4 and 1.5. Is there any correlation between the distribution of calcareous and siliceous sediments and red clays, and the main physiographic features? Where is most of the terrigenous sediment to be found?

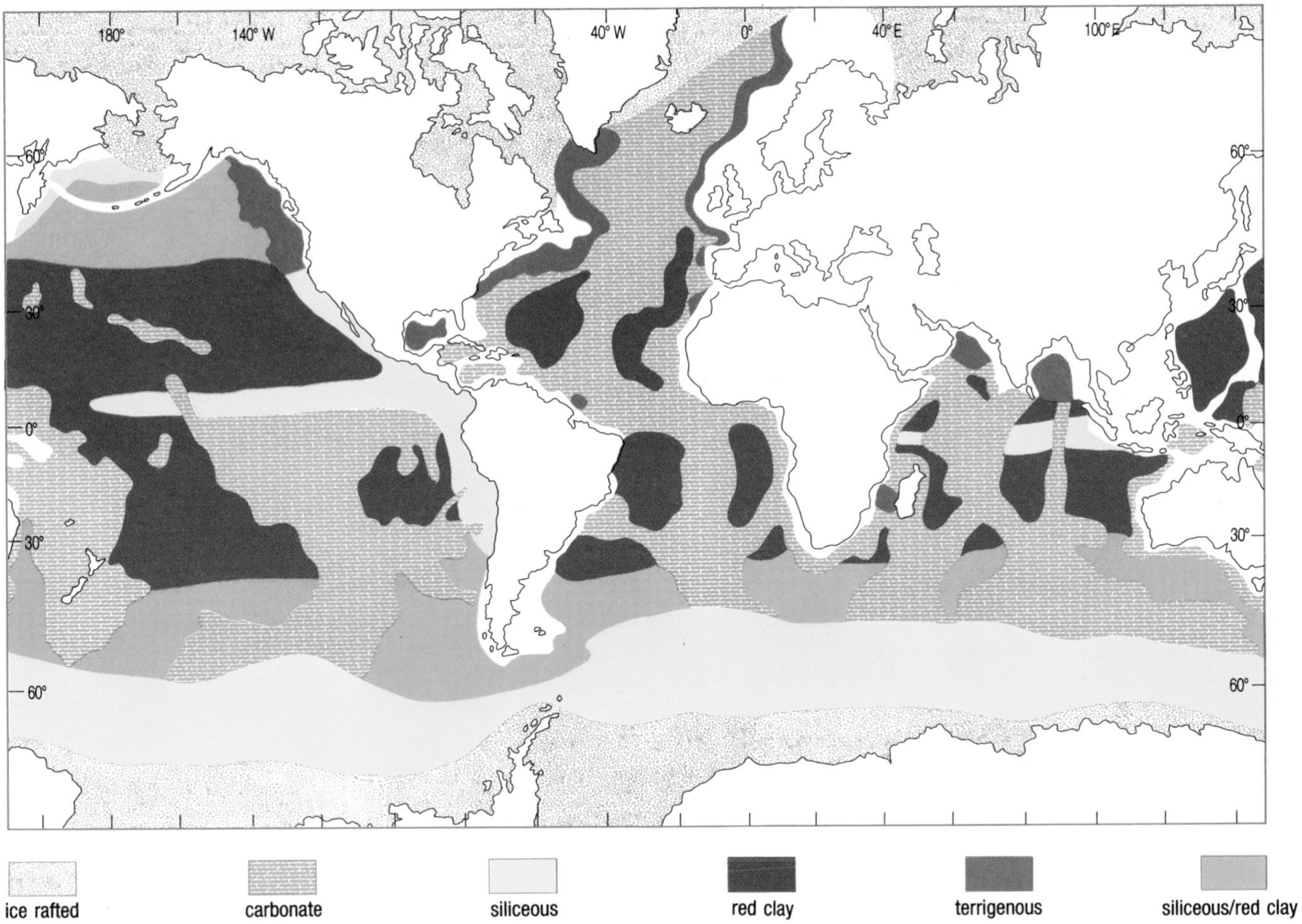

180° 140° W 40° W 0° 40° E 100° E

60° 30° 0° 30° 60°

ice rafted | carbonate | siliceous | red clay | terrigenous | siliceous/red clay

Figure 1.4 Distribution of dominant sediment types on the floor of the present-day oceans. Note that red clays are also terrigenous sediments.

Figure 1.5 Shaded relief map of the Earth's solid surface. In oceanic areas, the deeper the blue, the deeper the water.

At this point, we must emphasize that there is no such thing as a 'pure' terrigenous or biogenic sediment. For example, terrigenous dust is widely dispersed by winds and currents; biogenic sediments always contain material of non-biological origin; and most calcareous sediments contain some siliceous material, and vice versa. Conversely, terrigenous sediments are seldom without a biogenic component, however small; and it is worth mentioning that the predominance of red clays on abyssal plains is due to an *absence* of material which elsewhere dilutes the terrigenous component, rather than to a large flux of clay particles to the sea-bed from overlying waters.

There is a convenient 'rule of thumb' for broadly classifying deep-sea sediments, known as the '30% rule'. If the sediment contains more than 30% biogenic components, it is called a calcareous or siliceous ooze (depending on which biogenic component is dominant); if it contains less than 30%, it is a red clay.

Sediments deposited in the open ocean and beyond the influence of processes along continental margins are called **pelagic sediments**. This term encompasses both terrigenous and biogenic material that accumulates slowly from dispersed suspension in the oceanic water column. It thus excludes, for example, sediments that reach the abyssal plains from turbidity currents (see Chapter 4), or the sediments round oceanic islands. Two other sources of sediment to the oceans should be mentioned here, although they generally make up rather a small proportion of the whole. Volcanic eruptions eject ash and dust into the atmosphere, and this material may be carried great distances by the wind and washed out directly into the oceans by rain. The volcanogenic component of deep-sea sediments can be significant in the vicinity of active volcanoes and after major eruptions such as that of Krakatoa in 1886. There is also a small proportion of material of extra-terrestrial origin (Figure 1.6), the remains of meteorites destroyed in their passage through the Earth's atmosphere. The rate of accumulation of this meteoritic or cosmic dust in the deep sea is estimated to be in the range 0.1 to 1mm per million years.

Figures 1.4 and 1.5 are mere snapshots in terms of geological time. The ocean basins change their shape and size at rates of centimetres per year because of **sea-floor spreading** and **plate tectonics**; and global climatic regimes fluctuate on time-scales of 10^4 to 10^7 years. The oldest ocean floor was formed about 160Ma (million years) ago, but even that seemingly vast time-span represents only about one-thirtieth of the age of the Earth as a whole. Nonetheless, during that same time-span, the Earth has seen the break up of supercontinents and a transition from an essentially ice-free planet to one with polar ice-caps. The familiar current systems of the modern ocean did not exist 160Ma ago, and the distributions of temperature and salinity—not to mention the nature and variety of most marine life forms—were quite different from what they are now.

Does that summary imply anything about what you might find if you were to drill down through the sediments in those parts of Figure 1.4 that lie well away from the ocean ridges, i.e. where the sediments are thickest?

You would still be able to recognize the broad subdivisions of sediment *type*, that is, terrigenous versus biogenic, and calcareous versus siliceous.

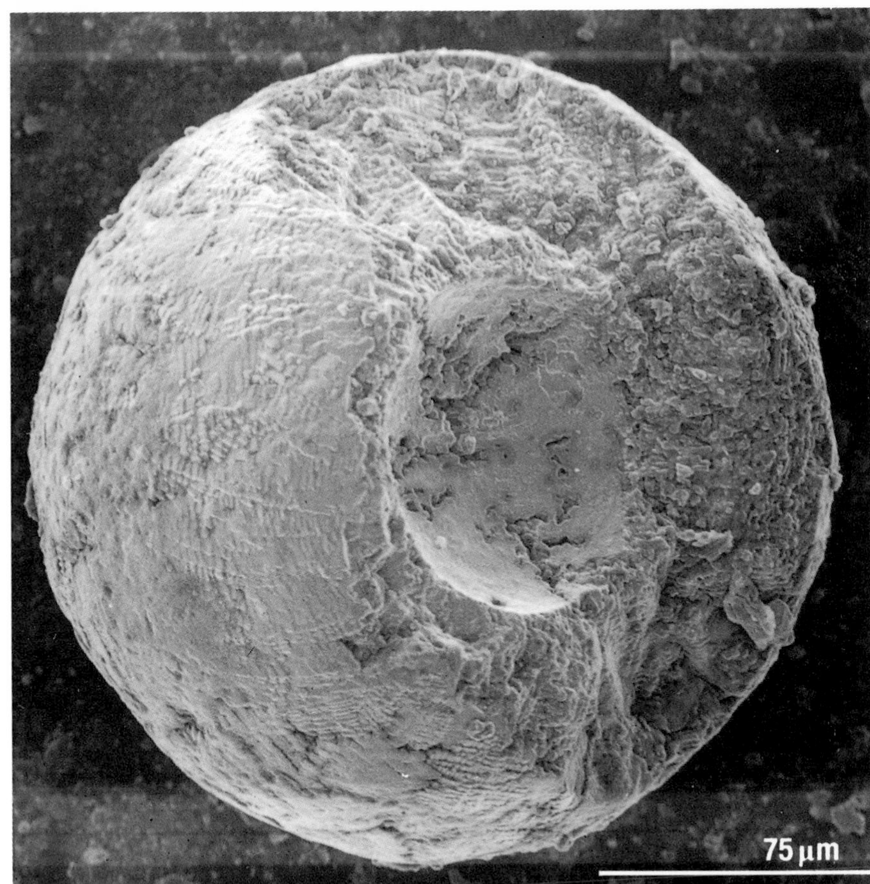

Figure 1.6 Scanning electron micrograph of a spherule of cosmic dust, typical of those found in deep-sea sediments. An iron-nickel core is surrounded by magnetite formed by oxidation during passage through the Earth's atmosphere. This spherule was found in red clay collected by HMS *Challenger* from the southern mid-Pacific. Cosmic spherules range from about 50 to 200 µm in diameter and from about 0.5 to 10 µg in weight. It has been estimated that 300×10^3 tonnes of cosmic (meteoritic) dust falls on the Earth's surface each year.

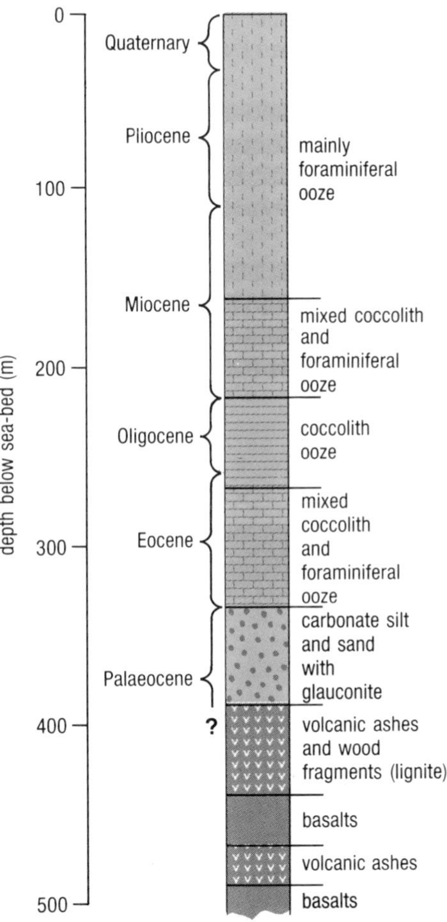

Figure 1.7 The sequence of sediments and volcanic rocks sampled at DSDP site 214 on the Ninety-East Ridge (Indian Ocean) at about 11°S, water depth 1665 m. *IMPORTANT*: the sequence is presented merely to demonstrate that the type of sediment deposited at a single location may change greatly with time. You do *not* need to recognize or understand the details at this stage; they will be explained later. For geological ages, see the Appendix. DSDP = Deep Sea Drilling Project, now superseded by Ocean Drilling Programme (ODP).

But those categories would almost certainly change with depth in the sequence (Figure 1.7); and secondly, within those categories, the proportions and types of non-biological components would probably be different. The species composition of the biogenic components would also be different—the more so, the deeper you drilled.

Equally important is the fact that while sediment sequences such as the one in Figure 1.7 accumulated on top of a particular piece of oceanic crust, that piece of crust will not be in the same geographical position as it was when first formed. New crust is formed at oceanic ridges and moves away from them, so that after even a few tens of millions of years it may be hundreds of kilometres from its place of origin.

Moreover, the sediments deposited on that piece of crust would no longer have their original chemical and mineralogical characteristics. Reactions between seawater and sedimentary particles begin during the comparatively short space of time taken to reach the sea-bed (rarely more than a few weeks). Reactions continue both at the sea-bed and within the growing sediment pile that accumulates as sediments continue to be deposited, because there is still plenty of seawater left in the pore spaces between individual sediment particles (see Chapter 5).

Finally, and perhaps most importantly, the record will be incomplete, because sedimentation is rarely continuous at any site. Episodes of erosion or non-deposition are common, as patterns of currents change and evolve.

1.1.1 PELAGIC BIOGENIC SEDIMENTS

Viewed through a microscope, biogenic sediments are seen to consist of a wonderful variety of delicate and intricate structures, mostly the skeletal remains of marine phytoplankton and zooplankton. As the lifespan of most of these planktonic organisms is only about a week or two, there is a slow but continuous 'rain' of their remains down through the water column to build up successive layers of sediment. As you will see shortly, the occurrence of the remains of any one particular type of planktonic organism depends upon a number of local factors such as water chemistry and depth, and the extent of primary production in the surface ocean waters. Because of this, the presence of these microfossils in ancient deep-sea sediments can be used to determine what the water depths and surface productivity were like during the geological past.

Carbonate sediments are composed principally of the skeletal remains of **coccolithophores**, **Foraminifera** and **pteropods**. Siliceous remains come mostly from **diatoms** and **Radiolaria**. The hard parts of these organisms vary a great deal in size, shape and chemical stability, and these factors control their preservation potential in deep-sea sediments.

Coccoliths are minute plates of **calcite** (the more common form of calcium carbonate, $CaCO_3$), usually less than 10μm in size (μm = micrometre (micron) = 10^{-6}m), with which the phytoplanktonic (algal) coccolithophores envelop themselves to form coccospheres (Figure 1.8). When the algal cell dies, the coccosphere disintegrates easily, releasing the individual plates—the coccoliths—into suspension. Thus, it is the coccoliths rather than the whole coccospheres that are preserved in sediments. Plates are also shed from the coccospheres as the algae grow, so coccolithophores contribute to deep-sea carbonate sediments before the death of the organism, as well as afterwards. Each coccolith has an organic membranous covering which inhibits dissolution of the calcite and enhances preservation. The white cliffs of Dover are formed of chalk that is mostly composed of coccoliths.

Because of their exceptionally small size, coccoliths are referred to as **nanofossils** (from the Greek word 'nanos', meaning dwarf) and carbonate sediments particularly rich in coccoliths are known as **nanofossil oozes** (or simply as coccolith oozes or nano-oozes).

Figure 1.8 Examples of coccolithophores.
(a) *Coccolithus huxleyi.* Both complete and disaggregated coccospheres; coccosphere diameter *c.* 6μm.

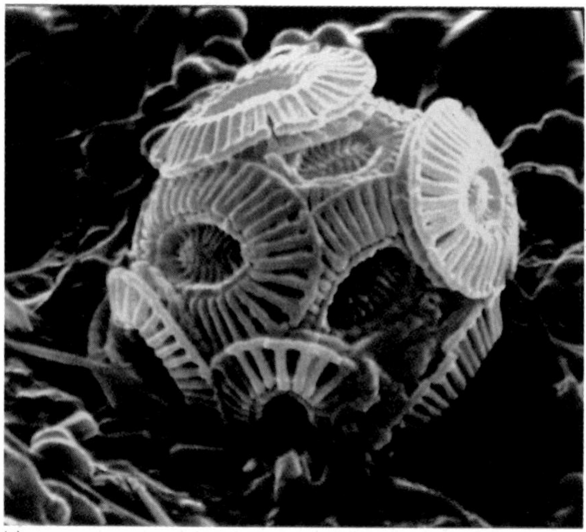

(a)

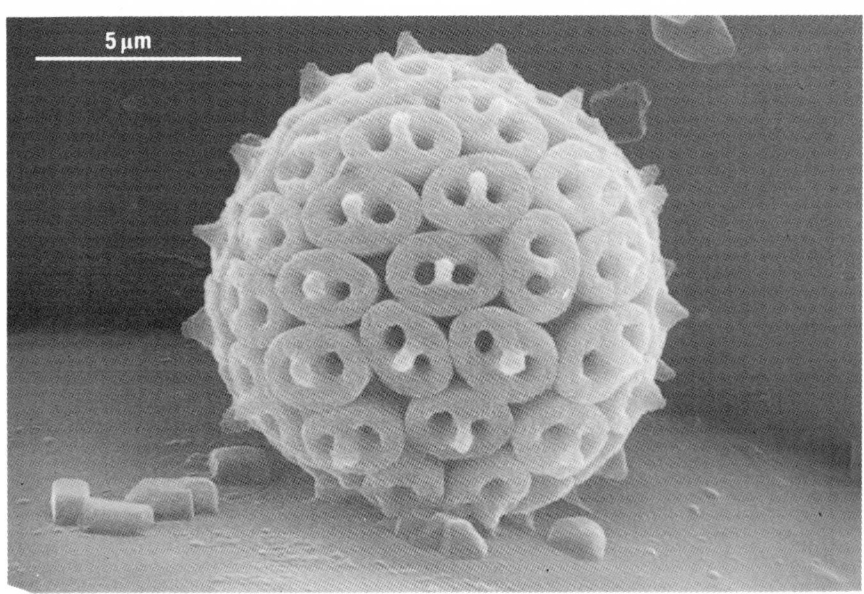

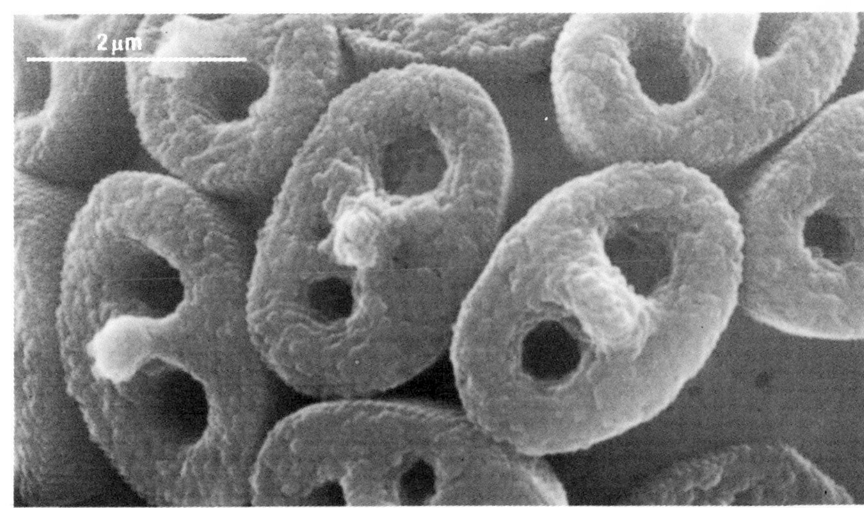

(b)

Figure 1.8 (b) *Homozygosphaera ponticulifera*. Both complete and disaggregated coccospheres; scale bars 5 μm and 2 μm.

(c) *Gephyrocapsa ornata*. Complete coccosphere; scale bar 5 μm.

(d) *Scyphosphaera apsteinii*. Complete coccosphere; scale bar 20 μm.

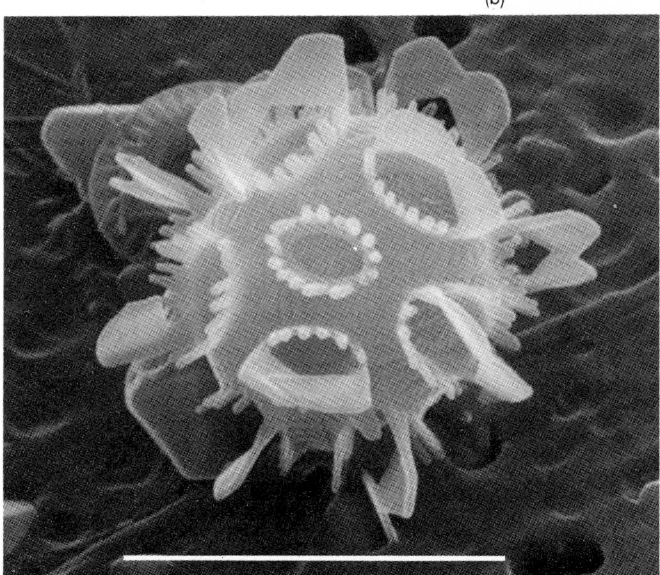

(c)

(a)

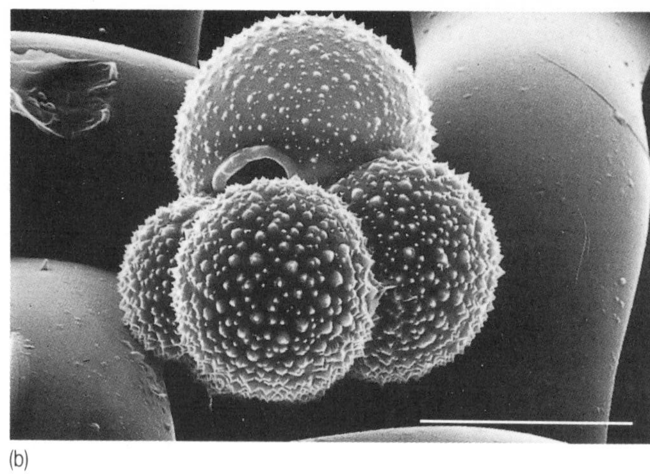

(b)

Figure 1.9 (a) Foraminifera from washed calcareous ooze. See also Figure 1.2(b).

(b) *Globigerina* sp. Scale bar 50 μm. The background is the mesh of the sampling net.

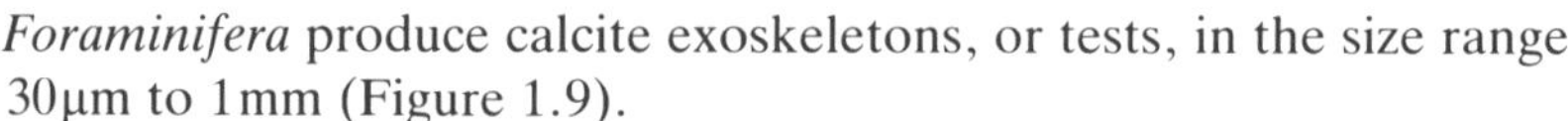

Foraminifera produce calcite exoskeletons, or tests, in the size range 30 μm to 1 mm (Figure 1.9).

Most zooplanktonic Foraminifera live in the top 1000 m of the water column. Surface and near-surface species tend to have a more spiny shape, which increases the ratio of surface area to volume and thereby aids a floating mode of life, as well as deterring predators. However, an abundance of spines also increases the surface area of a test, making it more prone to dissolution after death. Deeper-water species, which usually lack spines, are more commonly preserved. Because of their relatively large size, abundant Foraminifera give a sandy texture to the pelagic sediments in which they occur. There are also both shallow-water and deep-water forms of bottom-living or **benthic** Foraminifera, and in some regions these can be abundant.

Figure 1.10 Pteropods. The larger shells are about 3 mm long. See also Figure 1.2(c).

Pteropods are planktonic gastropod molluscs with thin shells up to a centimetre long (Figure 1.10). At the present day, most species are restricted to tropical and sub-tropical oceanic areas. Their shells are composed of **aragonite** (a variety of calcium carbonate, $CaCO_3$, that is more soluble than calcite), so they are more easily dissolved than calcitic plankton shells and are not found in ocean sediments where water depths exceed about 2–3 km.

Diatoms are unicellular algae ranging in size from a few μm to around 200 μm (Figure 1.11). They secrete shells of amorphous hydrated silica, the formula of which is commonly written as SiO_2nH_2O (and it is sometimes called opaline silica or opal). For brevity, we shall use SiO_2. Both planktonic and shallow-water benthic diatoms occur, but the planktonic species have thinner tests and are more prone to dissolution. The remains of planktonic diatoms often dominate the siliceous sediments found at high latitudes (Figure 1.4).

QUESTION 1.2 (a) Why are benthic diatoms restricted to shallow water?

(b) Can you suggest the maximum depth at which you might expect to encounter benthic diatoms?

Radiolaria are quite large zooplanktonic organisms (Figure 1.12), usually between 50 μm and 300 μm or more in size. They also have skeletons of silica and are the dominant biogenic component of siliceous sediments

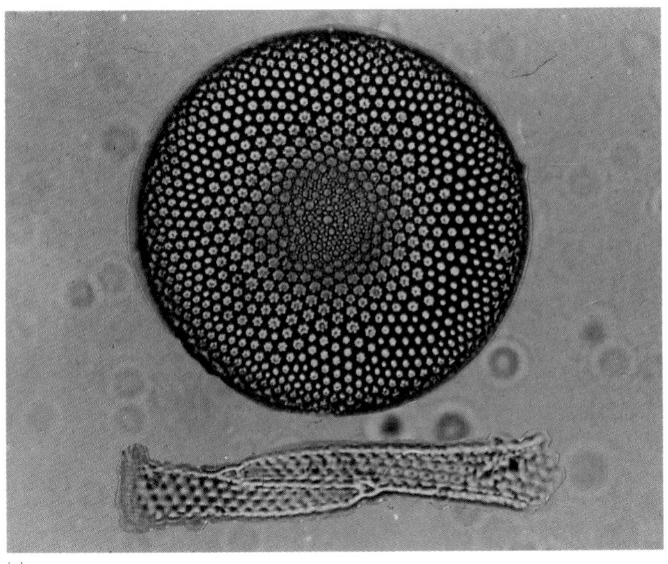

(a)

(b)

Figure 1.11 (a) Discoid (centric) and spindle-shaped diatoms. Width of field about 50 μm.

(b) Centric diatom. Scale bar 20 μm.

found at low latitudes. As with the Foraminifera, both surface and deeper-water species occur.

QUESTION 1.3 By analogy with Foraminifera, how would you expect the shape and preservation potential of surface forms of Radiolaria to differ from those of deeper-water forms?

It is important always to keep in mind that the distribution of biogenic sediments (Figure 1.4) is determined partly by the extent of biological productivity of the plankton in surface waters; and partly by the extent to which the skeletal remains are dissolved in the water column and at the sea-bed—and that in turn depends upon the chemical properties of seawater (see Chapter 3).

Figure 1.12 (a) Radiolaria from washed siliceous ooze. Individual tests *c.* 100 μm across. See also Figure 1.2(a).

(b) *Hexacontium* sp. Scale bar 50 μm. The background is the mesh of the sampling net.

(a)

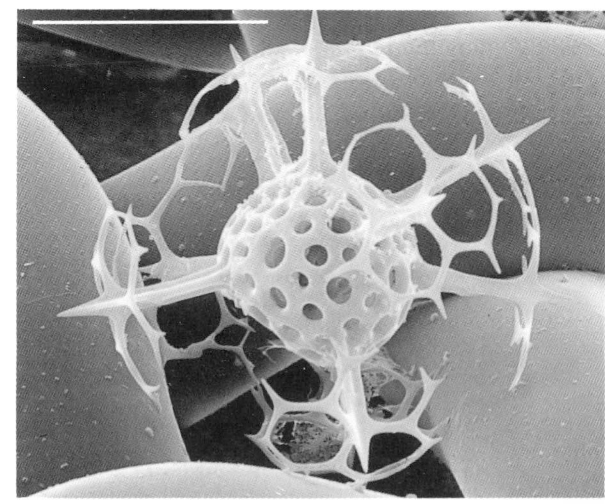

(b)

1.1.2 PELAGIC TERRIGENOUS SEDIMENTS

Nearly all terrigenous sediments in the pelagic environment are composed of material of the smallest grain sizes. There are two ways in which coarser-grained material can reach the pelagic environment. One is via turbidity currents and other gravity flows (see Chapter 4). The other way is through ice-rafting, that is, glacial material being shed by icebergs drifting into the open ocean and melting. Ice-rafted boulders, pebbles and sand may be found among pelagic sediments up to several hundred kilometres from the glaciers or ice-shelves from which they originally 'calved'.

The wind is obviously a powerful means of transporting fine material directly to the open oceans. Indeed, it is important to emphasize that pelagic clays deposited in the open oceans are predominantly of wind-blown (aeolian) origin and mostly less than about 20μm in size. The regions most likely to generate wind-blown dusts are the low-latitude belts influenced by persistent Trade Winds and low rainfall. The total amount of wind-blown dust delivered to the oceans annually is of the order of 10^8tonnes. This is very small compared with the sediment load supplied by rivers, which is about 1.5×10^{10}tonnes per year. Some river-borne sediment does reach the abyssal plains (see Chapter 4) but nearly all of it is deposited along continental margins; that includes clay minerals which are largely removed from suspension by **flocculation** in river mouths and estuaries. So, the proportion of river-borne terrigenous sediment in pelagic clays can be reduced relative to the aeolian contribution.

The principal components of wind-blown dusts are quartz and clay minerals, and these are being continually supplied to the oceans. More intermittent and much more spectacular is the supply of volcanogenic material to the pelagic environment though, on a global scale, the proportion of such material in deep-sea sediments is small (Section 1.1). Major eruptions can eject large quantities of volcanic ash and dust to heights of 15 to 50km, where the smallest particles, 1μm or less, can remain suspended for many months. During that time, they can be carried several times round the world by high altitude winds, causing unusual weather conditions and spectacular sunsets. Material between about 1μm and 20μm in size will rarely be projected to heights above 10km, and is deposited in a matter of days or weeks and within several hundred to a few thousand kilometres from the eruption. The result is distinctive layers of volcanic ash, which can be useful in the correlation of pelagic sediment sequences from widely separated locations.

Figure 1.13 Scanning electron micrograph of a fly-ash particle embedded in biogenic debris and clay minerals, collected in a sediment trap on the floor of the Sargasso Sea. Scale bar = 3μm.

There is a growing number of wind-borne anthropogenic contributions to the deep sea: some benign, others less so. They include dust from power stations and cement works (Figure 1.13); particles of more persistent plastics; PCBs (polychlorinated biphenyls from the plastics and electrical industries); lead compounds (mainly from motor vehicles); radionuclides; and various products from waste incineration at sea. The production of these materials is limited to the past hundred years or less, so their appearance in sediments can be used as a time-marker and in some cases also as a tracer for the movement of material through the oceans, on its way to the sea-bed.

The pelagic clays or so-called red clays (Figure 1.3) are found over large areas of the deep ocean floor, especially in the Pacific (Figure 1.4). They comprise a number of different types of clay minerals and, as we said earlier, owe their existence as a recognizable sediment type to the *absence* of other sediment components rather than to an abundance of clay minerals as such. Thus, pelagic clays are least diluted in the deepest parts of the ocean basins, and away from areas of high surface productivity. In these regions, not only is there a lack of biogenic material sinking from the surface, but also the chemistry of the bottom waters is such that any calcareous shell debris is dissolved (see Chapter 3). These regions are also well away from continental margins and so the clays are less diluted by other terrigenous sediments. Their colour is actually brown, rather than red, and results from oxidation of iron in the sediments, which have very low sedimentation rates and experience prolonged exposure to oxygenated bottom waters on the sea-floor.

Clay minerals are hydrated aluminosilicate minerals that occur as thin flakes, generally less than 2μm across. There are four main types of clay minerals in pelagic sediments: kaolinite, chlorite, illite and montmorillonite, each of which is formed in a different weathering environment. Their relative proportions in a pelagic clay vary according to the prevailing climatic and geological conditions in the source region and along the transport pathways, and according to the mixing processes that occur in the oceans.

Kaolinite is formed by the extreme chemical weathering of silicate minerals, especially feldspars, and is most abundant in low latitudes.

Chlorite occurs in both igneous and metamorphic rocks of the continents, but is destroyed by the chemical weathering that predominates in low latitudes. In general, therefore, chlorite is abundant in pelagic clays in high latitude environments, where physical weathering predominates and chlorite is released to the oceans in an unaltered state.

Illite is the most widespread clay mineral, but is more abundant in the Northern Hemisphere, where it may contribute up to 70% of the clay minerals in a sediment. Illites form under a variety of conditions and are not characteristic of any particular latitude belt; so, whether they dominate the sediment or not depends on the degree of dilution by other clay minerals.

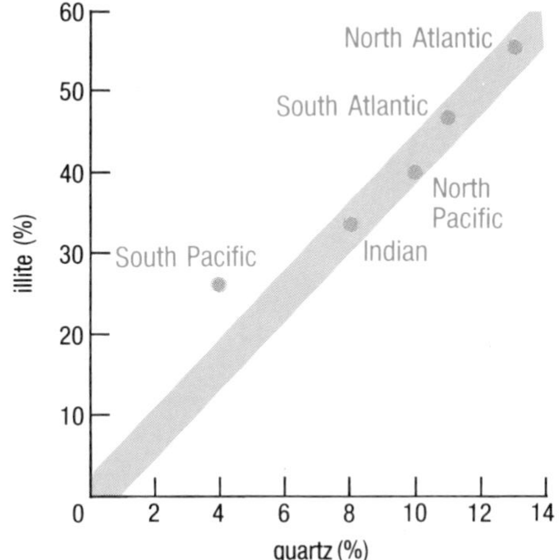

Figure 1.14 The relationship between the average concentrations of quartz and illite in the <2μm size fractions of sediments of the major ocean basins.

QUESTION 1.4 How well does Figure 1.14 confirm that last statement?

Montmorillonite is an alteration product of volcanogenic material, both in continental areas and in the marine environment. Much of the montmorillonite in deep-sea sediments is produced by the 'weathering' of volcanic ash actually on the sea-floor. Thus, strictly speaking, it is not always a terrigenous mineral, because some of it is formed *in situ* (see Section 5.2). It has been estimated that a layer of volcanic ash in a deep-sea sediment sequence can be totally altered to montmorillonite-rich pelagic clay in about 20Ma.

1.2 SEDIMENTS AND SEAWATER

The oceans receive a great variety of both solid and dissolved material from different sources. Deep-sea sediments are dominated by three types of components: calcium carbonate, silica and clay minerals (Figure 1.4). Relative amounts of these components in the sediments (Table 1.1) are significantly different from what they are in the particulate material that sinks from the surface. This is dominated by organic matter making up the body tissues of marine plants and animals, only *some* of which form skeletons of calcium carbonate or silica. Hardly any of the organic matter survives to reach the sea-bed—most of it is consumed and recycled in the top kilometre of the water column. Varying proportions of the skeletal materials are also dissolved before they can accumulate in the sediments.

By contrast, clay minerals and other inorganic solid products of terrestrial weathering are quantitatively almost insignificant in surface waters. However, they are relatively little affected by their passage through the water column, which is why they are so abundant in deep-sea sediments (Table 1.1).

We have seen that calcium carbonate and silica are being extracted from solution in seawater to form skeletons of plants and animals. But what about the other dissolved constituents? There is no evidence that the oceans are getting saltier with time, so these other constituents are somehow also being processed in the oceans and removed from solution. In the next Chapter, we examine the processes that cycle dissolved constituents within the main body of the oceans and ultimately remove them to the sediments.

1.3 SUMMARY OF CHAPTER 1

1 Deep-sea sediments can be classified broadly as terrigenous (land-derived) and biogenic (formed as the result of biological activity), with minor volcanogenic and cosmic contributions. Pelagic sediments include all those sediments deposited in the deep ocean basins beyond the influence of continental margin processes. Deep-sea sediments comprise material from more than one source. Factors such as submarine topography and climatic patterns influence the type of sediment that accumulates in a particular region.

2 The present-day distribution of deep-sea sediments is a reflection only of present-day climatic and current patterns and of the present configuration of ocean basins. The type of sediment deposited on a particular piece of ocean floor changes with time because of the way in which ocean basins, currents and climate change. The sequence of sediments recovered in a sediment core preserves a partial record of the changes that have occurred in the ocean above it.

3 Pelagic biogenic sediments are composed mostly of the remains of calcareous (carbonate) and siliceous planktonic organisms, principally coccolithophores, Foraminifera, pteropods, diatoms and Radiolaria. The preservation of these depends upon a number of factors such as water depth and chemistry, the shape of the skeletal remains, and the presence or absence of an organic membrane.

4 Wind-blown dusts comprise quartz and clay minerals and volcanogenic components. Coarse debris occurs in pelagic sediments only in high latitudes, deposited by ice-rafting. Pelagic clays comprise clay minerals derived from a number of different sources. They predominate in the deepest parts of the ocean basins where they are not diluted by biogenic material. Four main types of clay minerals are recognized, each characteristic of particular weathering regimes: kaolinite, chlorite, illite, and montmorillonite. Aeolian inputs are a major contribution to pelagic clays in many regions.

Now try the following questions to consolidate your understanding of this Chapter.

QUESTION 1.5 Suggest two reasons why sediments should be thicker near to continental margins than near mid-oceanic ridges.

QUESTION 1.6 It has been established by careful analysis that the ratio of the concentration of calcium to the total salinity of seawater is greater in deep than in surface water. Would you expect the ratio of dissolved silica (SiO_2) to total salinity to vary in the same way?

QUESTION 1.7 Explain whether you would expect to find chlorite or kaolinite in sediments: (a) of the equatorial Atlantic, (b) round Antarctica. (c) Why should illite be more common in sediments of the Northern than the Southern Hemisphere?

CHAPTER 2 CHEMICAL CYCLES IN THE OCEANS

'The quantity of the different elements in seawater is not proportional to the quantities of the different elements which river water pours into the sea ... but inversely proportional to the facility with which elements in seawater are made insoluble by general chemical or organochemical reactions in the sea.'

Georg Forchhammer (1865) *Phil. Trans. Roy. Soc.*

'When I think of the floor of the deep sea, the single, overwhelming fact that possesses my imagination is the accumulation of sediments. I see always the steady, unremitting, downward drift of materials from above, flake upon flake, layer upon layer – a drift that has continued for hundreds of millions of years, that will go on as long as there are seas and continents.... For the sediments are the materials of the most stupendous snowfall the Earth has ever seen....'

Rachael Carson, *The Sea Around Us.*

Most of the 92 naturally occurring elements have been measured or detected in seawater, and the remainder are likely to be found as more sensitive analytical techniques become available. The elements so far determined show a vast range of concentrations (Table 2.1).

Table 2.1 Average abundances of chemical elements in seawater.

Element		Concentration (mgl^{-1}) (i.e. parts per million, p.p.m.)	Some probable dissolved species	Total amount in the oceans (tonnes)
chlorine	Cl	1.95×10^4	Cl^-	2.57×10^{16}
sodium	Na	1.077×10^4	Na^+	1.42×10^{16}
magnesium	Mg	1.290×10^3	Mg^{2+}	1.71×10^{15}
sulphur	S	9.05×10^2	SO_4^{2-},$NaSO_4^-$	1.2×10^{15}
calcium	Ca	4.12×10^2	Ca^{2+}	5.45×10^{14}
potassium	K	3.80×10^2	K^+	5.02×10^{14}
bromine	Br	67	Br^-	8.86×10^{13}
carbon	C	28	HCO_3^-,CO_3^{2-},CO_2	3.7×10^{13}
nitrogen	N	11.5	N_2 gas,NO_3^-,NH_4^+	1.5×10^{13}
strontium	Sr	8	Sr^{2+}	1.06×10^{13}
oxygen	O	6	O_2 gas	7.93×10^{12}
boron	B	4.4	$B(OH)_3$,$B(OH)_4^-$,$H_2BO_3^-$	5.82×10^{12}
silicon	Si	2	$Si(OH)_4$	2.64×10^{12}
fluorine	F	1.3	F^-, MgF^+	1.72×10^{12}
argon	Ar	0.43	Ar gas	5.68×10^{11}
lithium	Li	0.18	Li^+	2.38×10^{11}
rubidium	Rb	0.12	Rb^+	1.59×10^{11}
phosphorus	P	6×10^{-2}	HPO_4^{-2},PO_4^{3-},$H_2PO_4^-$	7.93×10^{10}
iodine	I	6×10^{-2}	IO_3^-,I^-	7.93×10^{10}
barium	Ba	2×10^{-2}	Ba^{2+}	2.64×10^{10}
molybdenum	Mo	1×10^{-2}	MoO_4^{2-}	1.32×10^{10}
arsenic	As	3.7×10^{-3}	$HAsO_4^{2-}$,$H_2AsO_4^-$	4.89×10^9
uranium	U	3.2×10^{-3}	$UO_2(CO_3)_2^{4-}$	4.23×10^9
vanadium	V	2.5×10^{-3}	$H_2VO_4^-$,HVO_4^{2-}	3.31×10^9
titanium	Ti	1×10^{-3}	$Ti(OH)_4$	1.32×10^9

zinc	Zn	5×10^{-4}	$ZnOH^+, Zn^{2+}, ZnCO_3$	6.61×10^8
nickel	Ni	4.8×10^{-4}	$Ni^{2+}, NiCO_3, NiCl^+$	6.35×10^8
aluminium	Al	4×10^{-4}	$Al(OH)_4^-$	5.29×10^8
caesium	Cs	4×10^{-4}	Cs^+	5.29×10^8
chromium	Cr	3×10^{-4}	$Cr(OH)_3, CrO_4^{2-}$	3.97×10^8
antimony	Sb	2.4×10^{-4}	$Sb(OH)_6^-$	3.17×10^8
krypton	Kr	2×10^{-4}	Kr gas	2.64×10^8
selenium	Se	2×10^{-4}	SeO_3^{2-}, SeO_4^{2-}	2.64×10^8
neon	Ne	1.2×10^{-4}	Ne gas	1.59×10^8
manganese	Mn	1×10^{-4}	Mn^{2+}, $MnCl^+$	1.32×10^8
cadmium	Cd	1×10^{-4}	$CdCl_2$	1.32×10^8
copper	Cu	1×10^{-4}	$CuCO_3, CuOH^+, Cu^{2+}$	1.32×10^8
tungsten	W	1×10^{-4}	WO_4^{2-}	1.32×10^8
iron	Fe	5.5×10^{-5}	$Fe(OH)_2^+, Fe(OH)_4^-$	7.27×10^7
xenon	Xe	5×10^{-5}	Xe gas	6.61×10^7
zirconium	Zr	3×10^{-5}	$Zr(OH)_4$	3.97×10^7
bismuth	Bi	2×10^{-5}	$BiO^+, Bi(OH)_2^+$	2.64×10^7
niobium	Nb	1×10^{-5}	$Nb(OH)_6^-$	1.32×10^7
thallium	Tl	1×10^{-5}	Tl^+	1.32×10^7
thorium	Th	1×10^{-5}	$Th(OH)_4$	1.32×10^7
hafnium	Hf	7×10^{-6}	$Hf(OH)_5$	9.25×10^6
helium	He	6.8×10^{-6}	He gas	8.99×10^6
beryllium	Be	5.6×10^{-6}	$BeOH^+$	7.40×10^6
germanium	Ge	5×10^{-6}	$Ge(OH)_4, H_3GeO_4^-$	6.61×10^6
gold	Au	4×10^{-6}	$AuCl_2^-$	5.29×10^6
rhenium	Re	4×10^{-6}	ReO_4^-	5.29×10^6
cobalt	Co	3×10^{-6}	Co^{2+}	3.97×10^6
lanthanum	La	3×10^{-6}	$La(OH)_3$	3.97×10^6
neodymium	Nd	3×10^{-6}	$Nd(OH)_3$	3.97×10^6
lead	Pb	2×10^{-6}	$PbCO_3, Pb(CO_3)_2^{2-}$	2.64×10^6
silver	Ag	2×10^{-6}	$AgCl_2^-$	2.64×10^6
tantalum	Ta	2×10^{-6}	$Ta(OH)_5$	2.64×10^6
gallium	Ga	2×10^{-6}	$Ga(OH)_4^-$	2.64×10^6
yttrium	Y	1.3×10^{-6}	$Y(OH)_3$	1.73×10^6
mercury	Hg	1×10^{-6}	$HgCl_4^{2-}, HgCl_2$	1.32×10^7
cerium	Ce	1×10^{-6}	$Ce(OH)_3$	1.32×10^6
dysprosium	Dy	9×10^{-7}	$Dy(OH)_3$	1.19×10^6
erbium	Er	8×10^{-7}	$Er(OH)_3$	1.06×10^6
ytterbium	Yb	8×10^{-7}	$Yb(OH)_3$	1.06×10^6
gadolinium	Gd	7×10^{-7}	$Gd(OH)_3$	9.25×10^5
praseodymium	Pr	6×10^{-7}	$Pr(OH)_3$	7.93×10^5
scandium	Sc	6×10^{-7}	$Sc(OH)_3$	7.93×10^5
tin	Sn	6×10^{-7}	$SnO(OH)_3^-$	7.93×10^5
holmium	Ho	2×10^{-7}	$Ho(OH)_3$	2.64×10^5
lutetium	Lu	2×10^{-7}	$Lu(OH)$	2.64×10^5
thulium	Tm	2×10^{-7}	$Tm(OH)_3$	2.64×10^5
indium	In	1×10^{-7}	$In(OH)_2^+$	1.32×10^5
terbium	Tb	1×10^{-7}	$Tb(OH)_3$	1.32×10^5
palladium	Pd	5×10^{-8}	$Pd^{2+}, PdCl^+$	6.61×10^4
samarium	Sm	5×10^{-8}	$Sm(OH)_3$	6.61×10^4
tellurium	Te	1×10^{-8}	$Te(OH)_6$	1.32×10^4
europium	Eu	1×10^{-8}	$Eu(OH)_3$	1.32×10^4
radium	Ra	7×10^{-11}	Ra^{2+}	92.5
protactinium	Pa	5×10^{-11}	not known	66.1
radon	Rn	6×10^{-16}	Rn gas	7.93×10^{-4}

IMPORTANT: 1 Table 2.1 does not represent the last word on seawater composition. Even for the most abundant constituents, compilations from different sources differ in detail. For the rarer elements, many of the entries in Table 2.1 will be subject to revision, as analytical methods improve and more data become available.

2 Concentrations in Table 2.1 are by weight ($mg l^{-1}$ or p.p.m). While this is convenient for some purposes, for many others it is more useful to express concentrations in molar terms: one mole of any element (or compound) has a mass in grams equal to the atomic (or ionic, or molecular) mass of the element (or compound). Thus, a mole of calcium contains 40g; a mole of magnesium contains 24g; a mole of carbonate ion (CO_3^{2-}) contains $12 + (16 \times 3) = 60$g; and so on.

3 Concentrations of gases are given in $mg l^{-1}$ (p.p.m. by weight) in Table 2.1 and these are numerically not very different from the volumetric concentrations ($ml l^{-1}$) for oxygen, nitrogen, argon and some other gases. That is because of their low densities: 1.43, 1.23 and 0.77 $kg m^{-3}$ respectively. (So, to a first approximation, 1 m^3 weighs 1000g, 1 litre weighs 1 gram, and 1ml weighs 1mg.)

4 *Nitrogen and nitrate:* Dissolved nitrogen gas as N_2 is biologically almost inert and participates hardly at all in marine biological processes; only minute amounts of it are fixed (i.e. incorporated into living tissue) by micro-organisms. The total concentration of nitrogen in average ocean water is given as 11.5 p.p.m. in Table 2.1. Only about 0.5 p.p.m. of this total is in a form other than N_2 gas dissolved from the atmosphere, and it is this fraction which participates in marine biological cycles. It consists of fixed (i.e. chemically combined) nitrogen supplied to the oceans by rivers, mainly as nitrate produced by terrestrial nitrogen-fixing bacteria.

5 There is a wide variety of particulate matter in the oceans, and the distinction between what constitutes material truly in solution and what is particulate matter can present problems in the analysis of seawater. Filtration through a membrane having pores of diameter 0.45μm is a widely used procedure for separating dissolved from particulate fractions. This is satisfactory for most constituents, the exceptions being those occurring in hydrated forms that tend to coalesce into **colloidal particles**, which are nonetheless so small that they remain in suspension indefinitely (e.g. $Fe(OH)_3$).

2.1 BEHAVIOUR OF DISSOLVED CONSTITUENTS

Major constituents of seawater are those which occur in concentrations greater than about 1 part per million (1×10^{-6}) by weight. They account for over 99.9% of the **salinity** (S) of seawater, which for our purposes is the sum of all the dissolved salts in seawater. It approximates to 35 parts per thousand by weight throughout most of the oceans. Despite their relatively high concentrations, nitrogen and oxygen (Table 2.1) are not generally included among the major constituents because they are dissolved gases.

Minor and trace constituents make up the remainder of the elements in seawater. Although the distinction between the two is somewhat ill-defined, for convenience we here consider trace constituents to be those with concentrations of about 1 part per billion (1×10^{-9}) by weight, or less. On that basis, elements below about titanium in Table 2.1 are trace constituents.

Apart from the obvious distinction between major constituents and the rest in terms of mass, there is another reason for distinguishing between them. Salinity itself can vary from less than 33 to more than 37 parts per thousand throughout the oceans, depending on the extent to which freshwater is added by precipitation, run-off, and melting ice and snow, or removed by evaporation. But the ratio of the concentrations of most of the individual major dissolved constituents to total salinity remains practically constant. This **constancy of composition of seawater** is maintained because most of the major constituents exhibit **conservative** behaviour, that is, their concentrations in the seawater solution are not significantly changed by the biological and chemical reactions that take place within the main body of the oceans. Their concentrations can be changed only by mixing between different water masses of contrasted salinity. Nearly all the minor and trace constituents exhibit **non-conservative** behaviour: their concentrations are significantly changed by biological and chemical reactions in seawater.

From your reading of Chapter 1, which three of the major constituents must be exceptions to the generalization above, in that they are non-conservative?

They are carbon, calcium and silicon. In fact, calcium is so abundant in seawater that the ratio of its concentration to total salinity (the Ca:S ratio) is only very slightly greater in deep than in surface water, and this element departs only to a small degree from strictly conservative behaviour. The C:S ratio is more variable, and changes in the same direction as for calcium: there is more carbon in deep than in surface waters, and there is also some inter-ocean variability. The Si:S ratio varies greatly, both with depth and from place to place. Because of this strongly non-conservative behaviour, silicon (as dissolved silica, written SiO_2 for brevity) is generally not classified among the major constituents, even though the average concentration of Si is above the 1p.p.m. limit in Table 2.1.

A small number of minor and trace constituents also behave conservatively, but the great majority are non-conservative. Their concentrations can change considerably from place to place, but as those concentrations are so low (Table 2.1), such variations have no detectable effect either on total salinity or on the overall constancy of composition of seawater.

QUESTION 2.1 (a) How does Figure 2.1 suggest at first sight that nitrate and barium participate in biological cycles and sodium does not?

(b) Which of the profiles demonstrate(s) (i) conservative, (ii) non-conservative behaviour?

(c) Sodium is an example of an element that is essential to life, and is clearly not biologically 'inert'. How can this be reconciled with your answers to (a) and (b)?

At this point, it is necessary to emphasize that the distinction between conservative and non-conservative behaviour depends on the extent to which a constituent participates in biological or chemical reactions, *in relation to its overall concentration*. For example, small amounts of both magnesium and strontium can follow calcium into the carbonate skeletons of some marine organisms, and a minority of zooplankton (including some Radiolaria) can even secrete skeletons of celestite, strontium

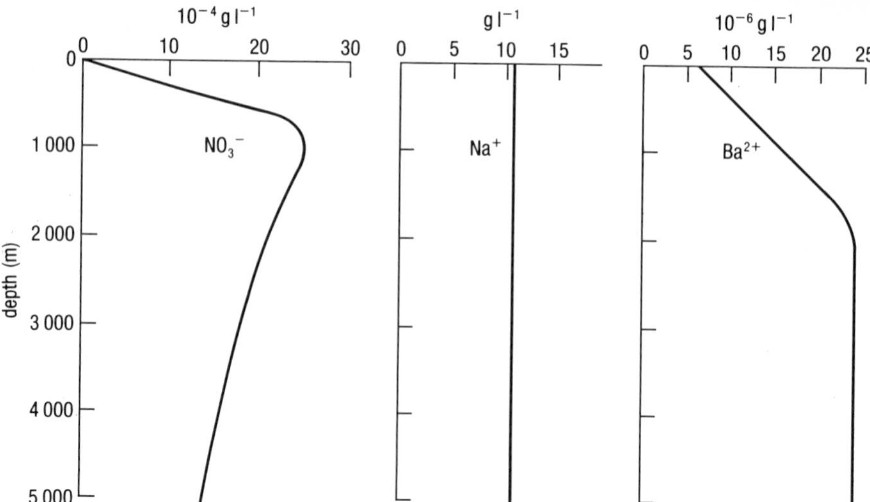

Figure 2.1 Concentration–depth profiles for three seawater constituents. Concentrations are normalized to a salinity of 35 parts per thousand. For discussion, see Question 2.1 and related text.

sulphate ($SrSO_4$). However, the amounts of magnesium and strontium involved in such transactions are so small in relation to their total quantity in the oceans that the effects are barely detected analytically. Magnesium is generally grouped among the conservative constituents, strontium among the non-conservative ones (but only just, as in the case of calcium).

2.1.1 THE STEADY-STATE OCEAN

The consensus among marine scientists is that the oceans are chemically in a **steady state**, at least for the major dissolved constituents of seawater (present in concentrations of more than 1p.p.m.) and probably for many of the minor and trace constituents as well. This means that the chemical budgets balance, such that the rate of supply of dissolved constituents equals the rate of removal. The available evidence suggests that the steady-state condition may be a characteristic feature of the oceans over periods of 10^8 years or longer. That is not to say that the ratios of average concentrations of individual constituents to total salinity would have been exactly the same in the oceans of, say, the Cretaceous (100Ma ago) as they are now; merely that they would have been generally similar.

The concept of the steady-state ocean allows us to define a mean oceanic **residence time**, which is given by:

$$\frac{\text{total mass of a substance dissolved in the oceans}}{\text{rate of supply (or removal) of the substance}}$$

Residence times of most elements are long compared with the average oceanic stirring or mixing time of about 500 years, so the majority of dissolved constituents should be uniformly distributed throughout the oceans. This is true for conservative constituents which have the longest residence times (more than 10^5 years) and whose removal from solution is accomplished mainly by inorganic processes. Non-conservative constituents are not uniformly distributed—concentrations vary with depth (*cf*. Figure 2.1) and from place to place, even though many have residence times that are much longer than the mean oceanic mixing time of 500 years. Until well into the 1980s, very few dissolved constituents were thought to have residence times of less than a few hundred years. As analytical methods improved and more data became available, however, residence times for some dissolved constituents were reduced to 100 years or less, when calculated on the basis defined above.

QUESTION 2.2 Explain why a dissolved constituent with an oceanic residence time of 100 years: (a) cannot be uniformly mixed throughout the oceans, (b) must be non-conservative in its behaviour.

Even the shortest residence times, however, are long compared with the lifespan of most marine organisms, so there is ample opportunity for dissolved constituents to participate in biological cycles several times over. All of this brings us to the relationships illustrated schematically in Figure 2.2(a) and summarized below:

1 Terrigenous sediments are supplied to the oceans by rivers, by winds, and from volcanic eruptions (meteorite infall is sufficiently small to be ignored here). Dissolved constituents are supplied mainly by rivers, but there are also inputs from volcanic gases and from **hydrothermal** systems, in which reactions between seawater and newly formed oceanic crust occur, especially at ocean ridge **spreading axes** (Figure 2.2(b)). There is a continuous exchange of gases between atmosphere and ocean across the air–sea interface.

2 Of the approximately 4×10^9 tonnes of dissolved material entering the oceans from rivers each year, about 10% consists of **cyclic salts**. Oceanic **aerosols**, formed by breaking waves and bursting bubbles, contain salts which are dispersed throughout the atmosphere and recycled back to the oceans by rain and rivers. Much of the chloride, nearly half of the sodium and significant amounts of other major dissolved constituents entering the oceans from rivers, are contributed as cyclic salts.

3 The majority of dissolved constituents participate in repeated cycles of biological activity that remove elements from solution and return them to it many times, before they end up in sediments. Some constituents (e.g. sulphate and magnesium) are also removed directly into oceanic crust during hydrothermal activity.

4 At the sea-bed itself, chemical exchange occurs between seawater and sediments as well as crustal rocks. Reactions that remove dissolved constituents from seawater are sometimes called reactions of **reverse weathering**, because some of them, especially the ones forming calcareous and siliceous sediments and some clay minerals (see Chapter 5), are essentially the reverse of the reactions that supply the necessary constituents during continental weathering.

5 The oceans are in a compositional steady state because the rates of supply of dissolved constituents from the various *sources* (items 1 and 2 above) are balanced by the rates of removal of those constituents into the *sinks* (items 3 and 4).

6 Ultimately, oceanic crust is carried back down into the Earth's mantle at **subduction zones**. Some of the overlying sediments (plus seawater trapped in them) are scraped off and accreted to continental margins, and some are carried into the mantle (Figure 2.2(c)). Both processes ensure that much of the material eventually finds its way back into the cycle of weathering and erosion at the Earth's surface and so eventually back into the oceans.

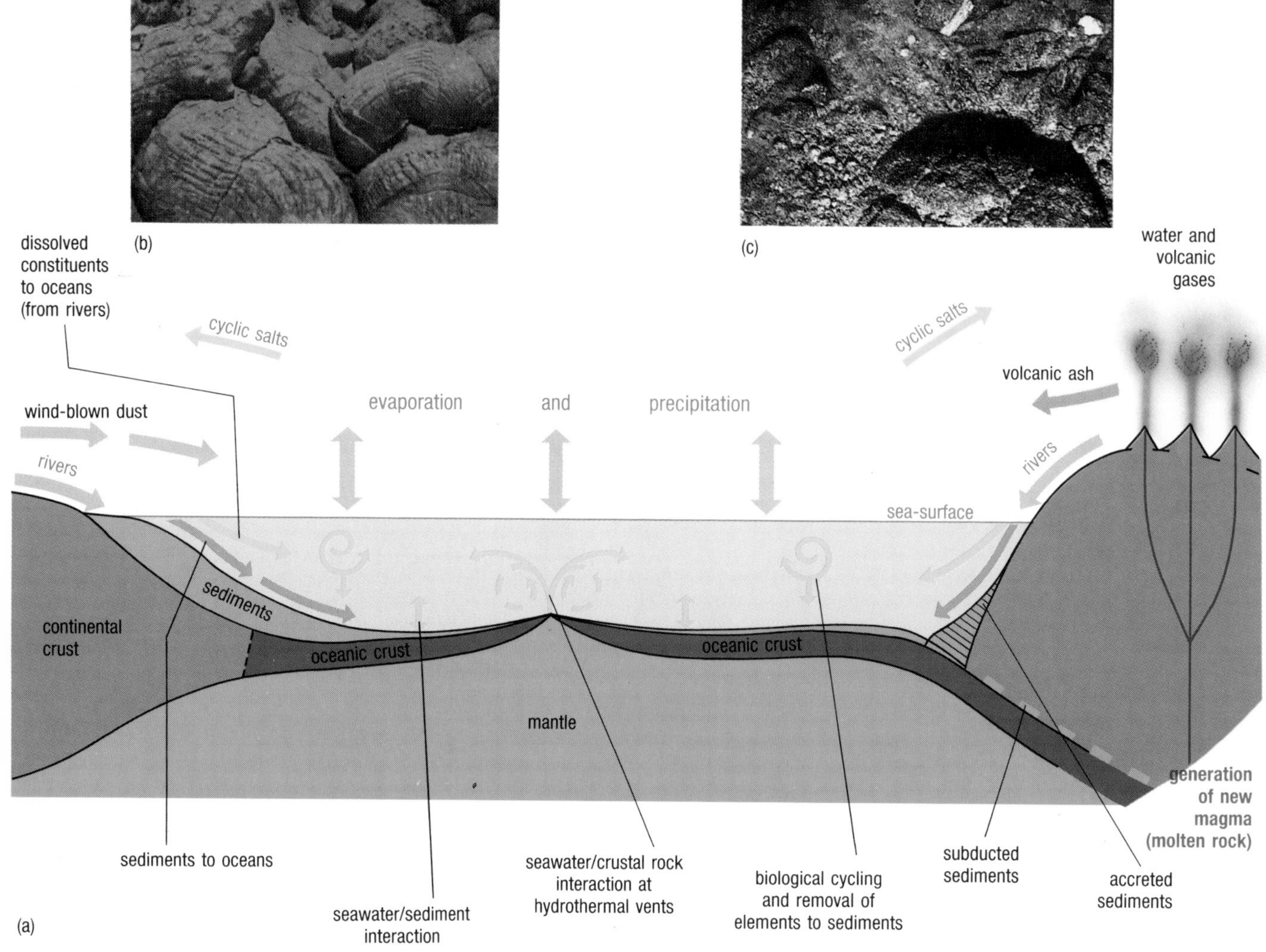

Figure 2.2 (a) Diagrammatic cross-section illustrating the role of oceanic cycles in the global cycling of elements through mantle, crust, rivers, atmosphere and oceans.

(b) Oceanic crust beneath the sediments consists largely of pillow lavas resulting from submarine volcanic activity at ocean ridge crests.

(c) Part of the east wall of the Marianas Trench, in the western Pacific, at a depth of 2380m. The slope of the sea-bed is about 45°. The rock is broken and shattered. This deformation is typical of the outer wall of oceanic trenches, beneath which oceanic rocks are deformed prior to subduction into the mantle.

We should not leave Figure 2.2 without mentioning **excess volatiles**. Mass balance calculations show that for several dissolved constituents of seawater, concentrations in crustal rocks are much too low for continental weathering to account for their abundance in the oceans. A good example is chloride, the most abundant constituent in the seawater solution (Table 2.1): its concentration in average crustal rock is 0.01%. Sulphur, boron and bromine are other major elements which are classed as excess volatiles and there are several minor and trace constituents also. All are found in volcanic gases, and the bulk of these excess volatiles may have been supplied to the oceans by volcanoes early in the Earth's history. They continue to be expelled in volcanic gases at the present time: it has been estimated that volcanic eruptions expel something like one million tonnes of chloride (as HCl) into the atmosphere each year (a minute amount compared with the total mass of chloride in the oceans, Table 2.1), but this is being augmented by increased inputs from anthropogenic sources.

In the past few decades and especially since the 1970s, it has become increasingly apparent that a major factor in the behaviour of many dissolved constituents, and hence in the regulation of seawater composition, is the oceanic particle cycle. Particles produced by biological processes in surface water sink into the deep ocean and participate in a variety of biogeochemical interactions on the way. Thus, as noted in Section 1.2, the relative proportion of individual components making up the sediments on the deep sea-floor (Figure 1.4) differs considerably from those in the original particles produced in surface waters.

2.2 THE BIOLOGICAL PARTICLE CYCLE

Most of the particulate matter falling from the surface layers of the ocean is produced initially by photosynthetic phytoplankton in the sunlit **photic zone**, which rarely extends more than 200m below the surface and generally much less. The phytoplankton are grazed by zooplankton that package most of their waste products into faecal pellets, which are in turn consumed and decomposed by other organisms, including bacteria. Thus, most of the organic matter is recycled in the surface layers, but a small proportion survives in the particles that sink out of the photic zone towards the sea-bed (Figure 2.3). As they fall through the water column, the organic matter in them not only provides food for successive populations of filter-feeders and other animals, so that it is re-packaged several times *en route* to the sediment, but also carries its own communities of microbial decomposer organisms.

The principal chemical elements making up the organic matter which forms the soft tissue of all organisms are: oxygen, hydrogen, carbon, nitrogen and phosphorus. The first three are abundant everywhere, but supplies of nitrogen and phosphorus in forms that are biologically utilizable are not always available, which is why these two elements are the two most important biological nutrients.

About three-quarters of the organic matter in the sinking particles that leave the photic zone is decomposed and recycled in the upper 500–1000m of the water column, i.e. above the main **thermocline**; and, on average, only about 1% of the sinking organic matter actually reaches the sediments. In other words, of the 5 units that contribute to the sinking particles in Figure 2.3, 3 or 4 units are recycled above the thermocline, and on average only about 0.05 unit survives to reach the sediments; and so organic-rich sediments are rare in the deep oceans. By contrast, the proportion of skeletal material increases as particles sink deeper because it is dissolved by chemical processes, which in general operate more slowly. *Thus, the mean composition of the particulate matter changes with depth such that the proportion of skeletal and inorganic components increases greatly*. Before looking at these processes in more detail, we first examine the nature of the particulate organic matter itself.

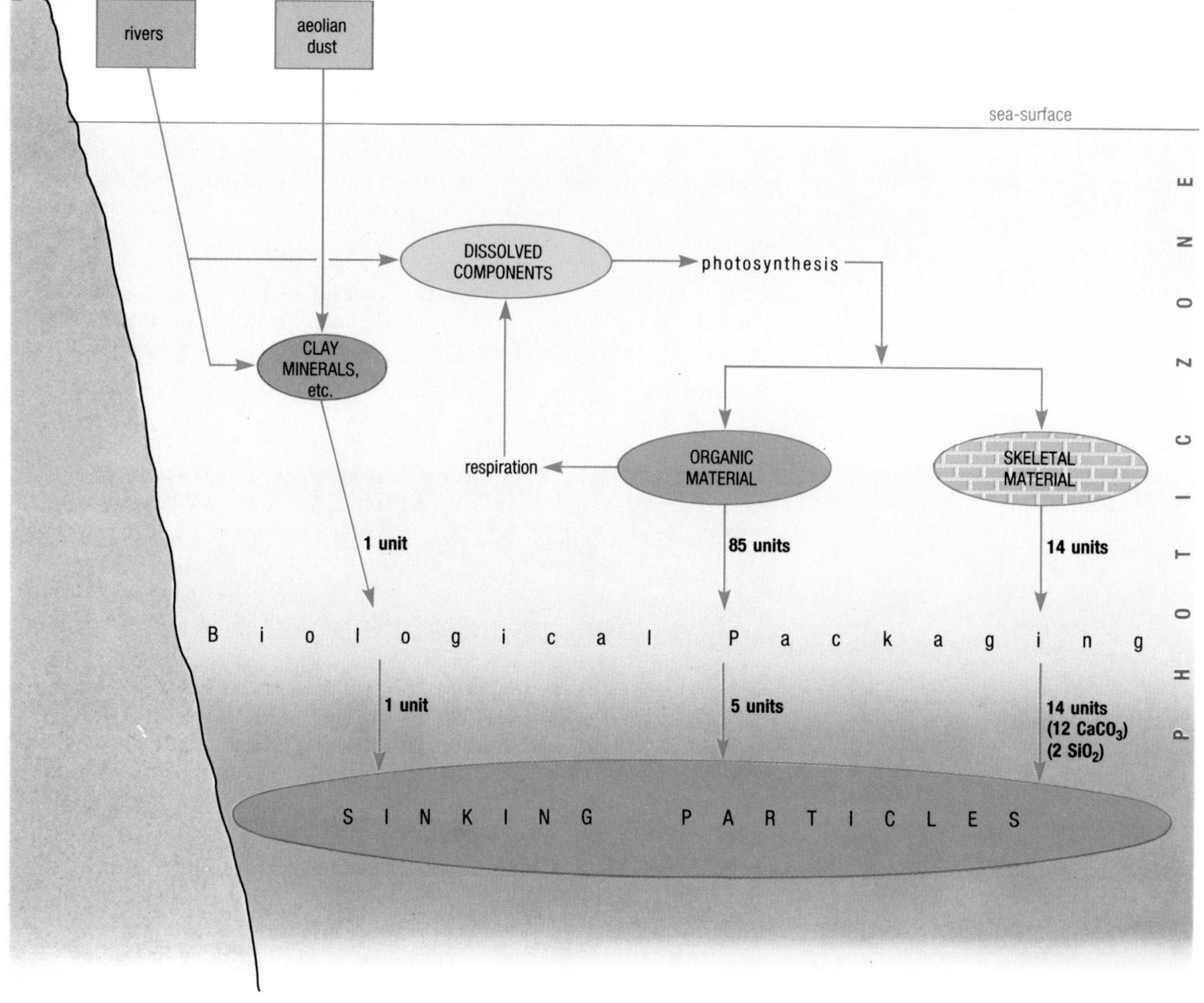

Figure 2.3 Simplified schematic diagram of biological particle formation in the oceans, showing approximate average make up of the particulate material. The high degree of recycling of organic matter in the photic zone ensures that on average it comprises only about one-quarter of the material sinking into deep water. (Note that of the 14 units of skeletal material, on average 12 units are $CaCO_3$, 2 units are SiO_2.)

2.2.1 PARTICULATE ORGANIC MATTER

Particulate matter in seawater (sometimes called the **seston**) can be approximately subdivided according to size—and, as already noted, in surface waters most of it is of biological origin (Figure 2.3). The smallest particles (less than 1 μm up to a few tens of μm) comprise bacteria and algal cells, and other fine organic detritus; coccoliths and diatom skeletons; and inorganic particles, especially clay minerals and insoluble hydrous compounds such as $Fe(OH)_3$. The size range from tens to a few hundreds of μm is represented by larger detritus and faecal pellets, the products of **biological aggregation** or packaging.

Finally, there are the easily visible (macroscopic) aggregates known as *marine snow* (or 'fluff'), consisting of detritus, living organisms (including bacteria) and some inorganic matter (mostly clay mineral particles). These aggregates are typically several millimetres across and in favourable circumstances they can attain dimensions of several centimetres.

Information about the nature and distribution of particulate organic matter in the water column comes from direct sampling of the water at various depths, using containers or filters, and from material collected in sediment traps—the oceanographers' equivalent of rain gauges, which they rather resemble in appearance. Sampling of marine snow presents particular problems because of the large size and great fragility of the aggregates, which may be broken up during collection, with the result that samples are not representative. Additional information has been obtained from direct observation by divers, during submersible operations, and from sea-bed photography.

Marine snow aggregates provide a fascinating and at first sight bewildering array of shapes and component materials, and you are not expected to understand all the details of Figures 2.4 and 2.5. The aggregates provide mini-ecosystems (microhabitats containing rich microbial communities and high nutrient concentrations), within which the processes of photosynthesis, decomposition and nutrient regeneration may occur at rates greater than those in surrounding waters. The chemical and biological properties of the particles change on time-scales of hours to days as the microbial communities undergo complex successional changes.

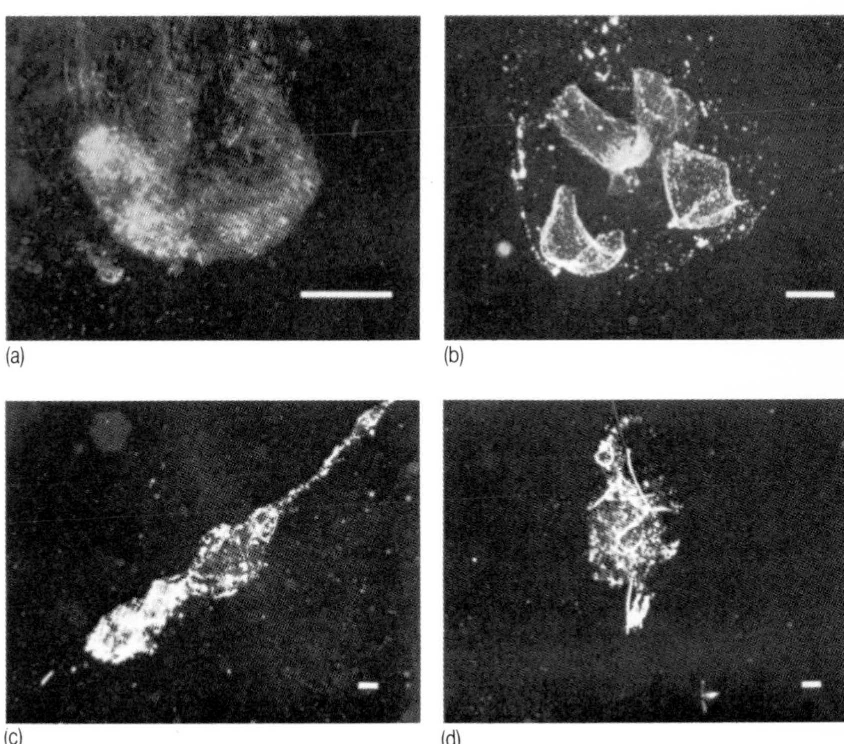

Figure 2.4 Examples of marine snow.

(a) Loosely associated aggregate of living, chain-forming diatoms. Scale = 1 cm.

(b) Abandoned filter nets of an appendicularian (small ascidian mollusc) surrounded by an envelope of particle-studded mucus.

(c) Typical comet-shaped aggregate.

(d) Irregularly shaped aggregate of unknown origin containing numerous macrocrustacean faecal pellets. Scale of (b)–(d) = 1 mm.

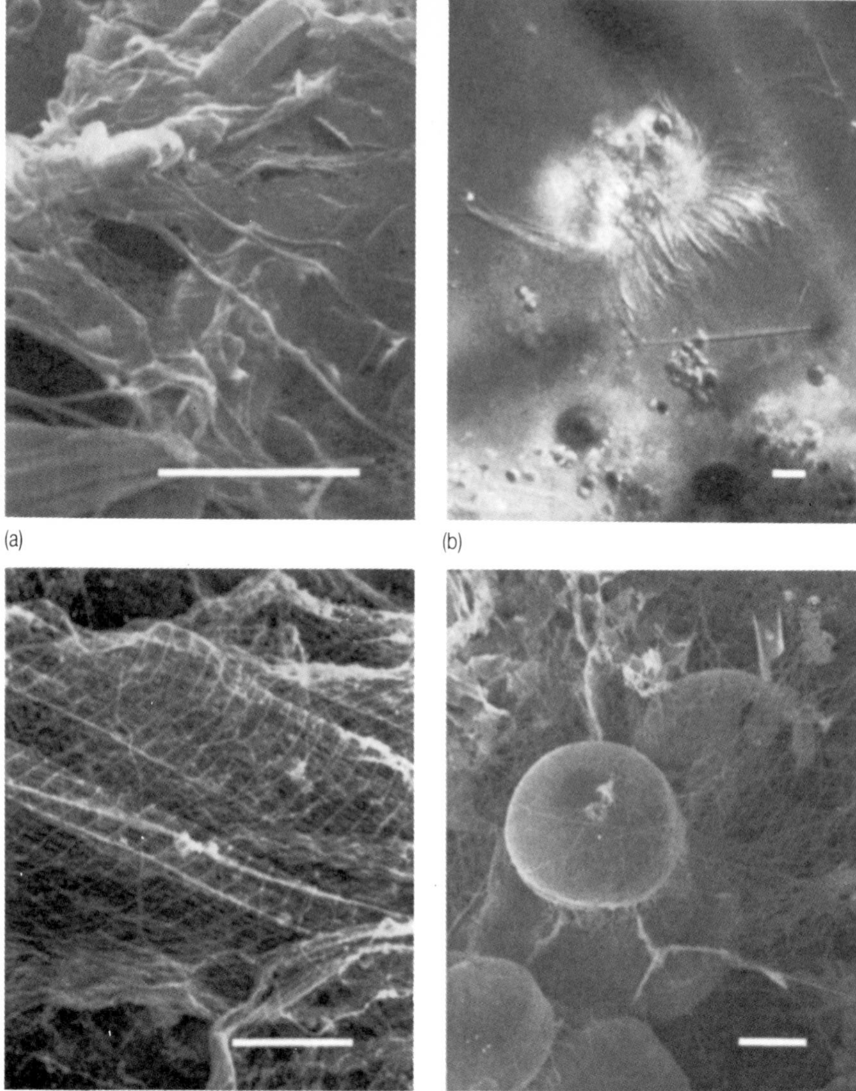

Figure 2.5 Micrographs of marine snow. Scale = 10 μm. SEM = Scanning Electron Micrograph.

(a) SEM of typical sheet mucus with diatom fragments.

(b) Ciliate on marine snow collected in a sediment trap at 1400 m off central Mexico.

(c) Typical filter web of an appendicularian 'house'.

(d) SEM of the centric diatom *Thalassiosira* sp. and its associated fibrils and mucus (sample from sediment trap at 50 m off California).

Figure 2.6 summarizes the formation and subsequent history of marine snow aggregates and shows how they can be formed either directly by living plants and animals (especially the mucus feeding webs of zooplankton) or by biological and physical aggregation of smaller particles.

What is likely to be the most important biological mechanism whereby small particles are aggregated together?

It is likely to be feeding, which aggregates small particles into larger faecal pellets. Of the physical mechanisms that can lead to aggregation, Brownian motion dominates the interactions of the smallest (μm-sized) particles, while differential settling and turbulence in the water column lead to collisions between particles and capture of small particles by large ones.

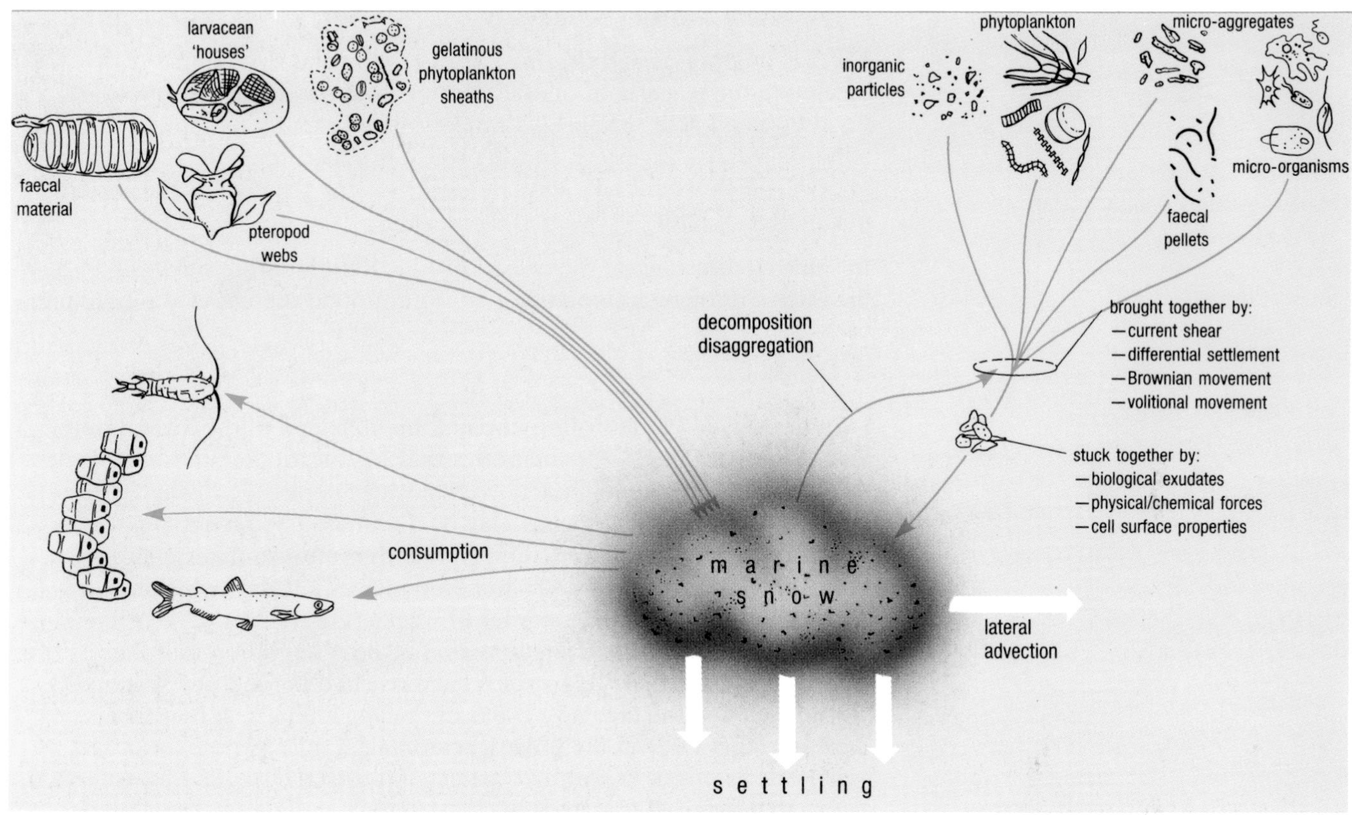

Figure 2.6 The formation and breakdown of marine snow. Aggregates are produced either directly in various forms and combinations by planktonic organisms; or indirectly, by the aggregation and adhesion of existing smaller particles.

QUESTION 2.3 Figure 2.6 illustrates three processes of breakdown of marine snow aggregates and two processes whereby it is removed from its place of formation. What are these processes?

As you might expect, marine snow is most abundant in surface waters, where production is high and component particles are plentiful. Local mid-water maxima observed near some continental margins are attributed to resuspension of aggregates on the shelf or lower down on the continental slope, by currents or internal waves; and their subsequent concentration on equal density (isopycnic) surfaces. The biological 'glues' that bind the aggregates are not especially strong, and marine snow must be continually disintegrating and re-forming, shedding some of its components and gaining new ones as it sinks towards the sea-bed and is disrupted by animals and by water movements.

Marine snow comprises the largest particles in the oceans and you might therefore suppose it to be the most important means of getting all kinds of particulate matter, both organic and inorganic, to the sea-bed. That is only partly true. Most marine snow is disaggregated and/or eaten and repackaged into faecal pellets in the upper 500–1000m of the water column, that is, above the permanent thermocline. Below about 1000m, faecal pellets (100–300μm) provide the main component of the sinking particle flux. Indeed, faecal biopackaging by zooplankton has been generally considered the principal reason why siliceous and calcareous oozes on the deep sea-floor lie beneath surface populations of the corresponding phytoplankton (*cf*. Figure 1.4). Faecal pellets not only sink relatively rapidly and so are not carried very far sideways; they also help to protect the skeletal remains from dissolution.

Vertical migrations and feeding habits of many marine animals can accelerate the transfer of organic matter from surface to deep waters. Food ingested near the surface can be carried rapidly to depths of several hundred metres, digested, and the waste products ejected. A depth interval which would take days or weeks to sink through is thus covered in a matter of hours.

In some circumstances, the rate of production of marine snow can exceed the rate of disaggregation and/or consumption in the upper water column.

What might those circumstances be?

Strong seasonal blooms of phytoplankton may lead to the formation of marine snow in such large quantities that significant amounts can reach the sea-bed more or less intact, even in water depths of 4000m or more. In some places, the aggregates may be dominated by phytoplankton (usually diatoms) that have clumped together into centimetre-sized 'flocs' with sinking velocities of a few hundred metres per day, which means they reach the sea-bed in a matter of weeks (Figure 2.7). One of the most interesting aspects of this phenomenon is the observation that the phytoplankton in these aggregates have reached the sea-bed without passing through the digestive tracts of grazing animals. It has been suggested that early in the growth season algal flocs are the most important means of exporting organic matter to the sea-bed from surface waters, while faecal transport becomes important later as zooplankton increase in abundance. In more general terms, marine snow is mostly found where there is high biological productivity, which is usually seasonal. In regions of lower productivity, faecal pellets make up most of the sinking particle flux.

The spectrum of particle sizes in the water column is sustained in a more or less steady state by the continuous exchange between larger sinking particles and smaller suspended (non-sinking) particles, through aggregation and disintegration. Smaller particles coagulate as they collide because of Brownian motion and/or are carried downwards by larger sinking particles which may then disaggregate and release them once

Figure 2.7 Photographs of the sea-bed at 4000m depth in the north-east Atlantic were taken with a time-lapse camera in 1983. Between 1 May and 15 June (a), there was little change (the mound in the foreground was made by an animal). During the next few weeks, heavy falls of marine snow (called 'fluff' by the scientists who took the pictures) all but obliterated the sea-bed, and by 14 July (b) only the top of the mound was still visible. The origin of the fluff was a phytoplankton bloom during May and early June.

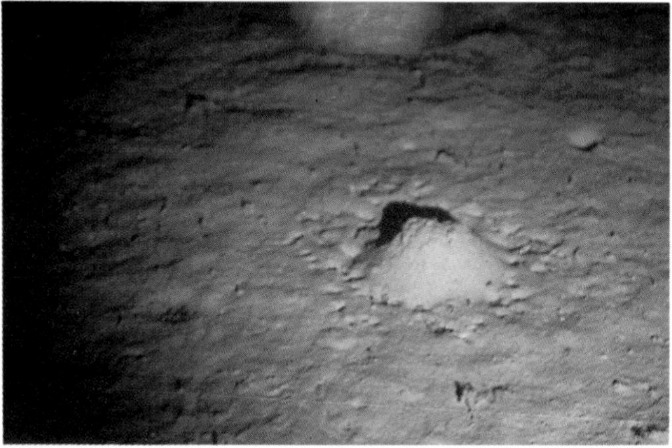

(a)

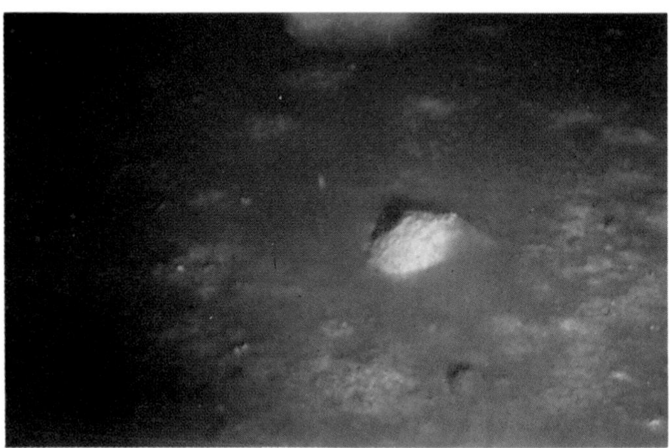

(b)

more. Thus, even the suspended particle population moves gradually downwards, being replenished by the production of new particulate material in surface waters. Settling rates of the smallest particles are so low that—even in the absence of turbulence in the water column—they would take a hundred years or more to sink to the sea-bed by themselves. However, their average lifetime in seawater is only about 7.5 years, which shows how effectively they are removed through collision and capture by larger sinking particles.

The size distribution of particulate matter in seawater is such that there is a more or less exponential increase in the numbers of particles, as particle diameter decreases. Volume for volume, therefore, there are many orders of magnitude more small particles than large ones: on average, of the order of 10^6–10^8 per litre or more for μm-sized particles, compared with about 1–10 per litre or less for marine 'snowflakes'. The oceanic 'soup' of particulate matter is actually rather thin, however. Marine snow rarely occupies more than 0.01% of the water column by volume (*c.* $10\,mg\,l^{-1}$ by weight) and commonly much less; while at the other end of the size range, concentrations of particles less than 5 μm in diameter generally average about $20–25\,\mu g\,l^{-1}$. Overall, the average concentration of particulate organic matter in seawater is equivalent to less than one-tenth of a milligram of carbon per litre ($0.05–0.1\,mg\,C\,l^{-1}$).

However, concentrations vary seasonally and from place to place in the oceans, because of variations in biological productivity. For example, the rather generalized map in Figure 2.8 distinguishes generally less productive **oligotrophic** regions of the open oceans from more productive **eutrophic** regions, on the basis of the organic carbon content in underlying sediments. The oligotrophic regions correspond roughly with the central parts of the oceanic gyres, and the level of primary productivity in these areas has been a matter of much discussion and debate. Data from satellite imagery and ship-based observations combine to suggest that a major part of open ocean production may come from localized transient blooms lasting a few days at most, rather than from a continuing low level of production over wide areas. The result would be intense local 'snowfalls' at the sea-bed, rather than a continuous light 'drizzle'; and the time-averaged rate of removal of carbon to the sea-floor might well not be the same in each case.

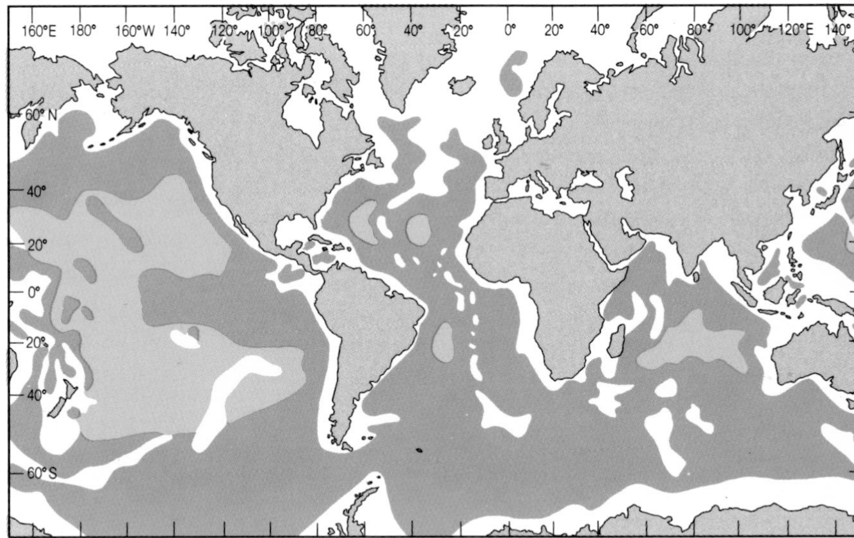

Figure 2.8 Division of the ocean floor into oligotrophic and eutrophic regions, according to the organic carbon content of surface sediments. Darker green = eutrophic areas (0.25–1.25%C); paler green = oligotrophic areas (<0.25%C). White areas represent topographic highs where there is little pelagic sedimentation.

A proper understanding of these processes is becoming increasingly important. The degree to which the oceans can mitigate the build up of greenhouse gases in the atmosphere depends partly on the rate at which atmospheric carbon dioxide can be sequestered by planktonic organisms and subsequently removed by the sinking and decomposition of particulate organic matter (and $CaCO_3$ in skeletal material).

Dissolved organic matter

The great majority of particles in suspension (<5μm) consist of free-living bacteria, subsisting partly on the smallest fragments of organic detritus, but mainly on dissolved organic compounds. These include organic acids, vitamins and sugars, mostly produced as metabolic by-products of the phytoplankton in particular. In addition to providing the main substrate for bacterial growth, this chemical 'conditioning' of the water by living algal cells can also influence seasonal successions of phytoplankton. Some species are unable to synthesize certain required organic compounds but can absorb them from their environment; and they can start to grow only when the water has been conditioned by those species capable of synthesizing and releasing the necessary compounds. Dissolved organic matter also includes compounds with very large relative molecular masses, in the region of 20000 or more. These are resistant to assimilation and bacterial decomposition and some estimates place their oceanic residence time at about 6000 years, orders of magnitude greater than that of any other organic matter in the oceans.

An important distinction between particulate and dissolved organic matter is that the dissolved material cannot sink into the deep ocean, but must be **advected** or mixed downwards. It can also be transported laterally over long distances from its sources by ocean currents before being broken down. Overall, the average concentration of dissolved organic matter in seawater is equivalent to about 0.5 to 1mgC l^{-1}, rather more than the average for particulate organic matter. There is a continual exchange of material between these two 'pools' of organic matter, as a result of biological activity.

Not all of the organic matter in seawater (whether particulate or dissolved) is produced there. Some comes from terrestrial sources, transported to the sea by rivers and by winds in particulate forms, and by rivers in dissolved forms. The latter include complex humic compounds as well as nutrients and are sometimes called **yellow substances**, for the distinctive colour they give to productive inland waters. Organic matter of terrestrial origin may be locally more abundant than organic matter of marine origin in coastal and shelf waters; but most of the particulate component is consumed or deposited there, and in global terms little survives to be exported to the deep sea. Particulate organic matter of terrestrial origin makes a small contribution to the annual aeolian supply to the oceans of *c.* 10^8tonnes (Section 1.1.2).

2.2.2 PARTICULATE ORGANIC MATTER AND THE NUTRIENT CYCLE

Sinking of particulate organic matter and the vertical movements of zooplankton and other animals feeding on smaller organisms and detritus combine to cause a progressive downward movement of nutrients out of the photic zone. Quantitatively, the three most important nutrients are fixed nitrogen (chiefly as nitrate, NO_3^-), phosphorus as phosphate (PO_4^{3-}) and dissolved silica (SiO_2 for brevity, but mainly as $Si(OH)_4$, Table 2.1).

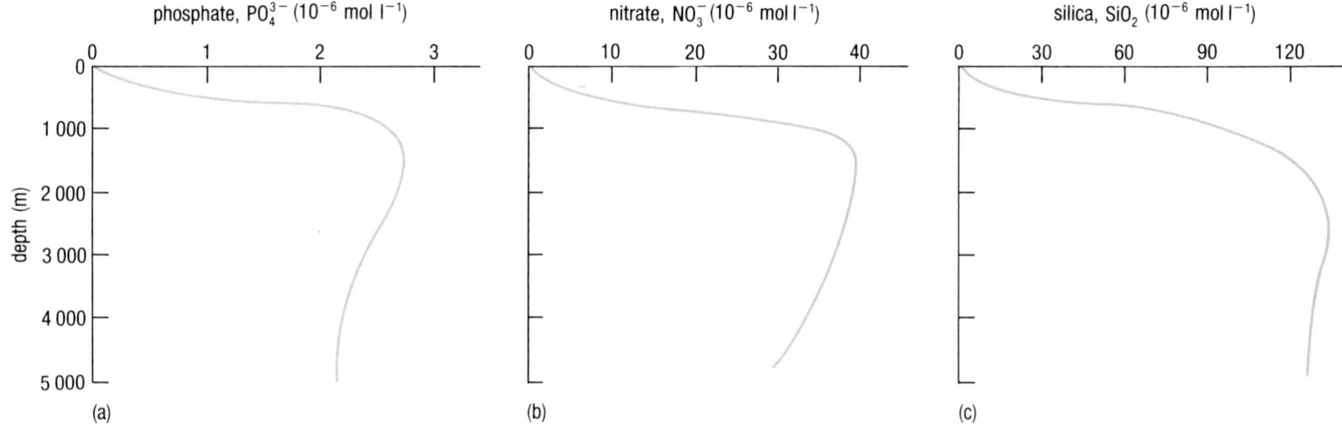

Figure 2.9 Typical concentration–depth profiles for (a) phosphate, (b) nitrate, (c) silica. All concentrations are in molar terms.

They are most heavily utilized in the photic zone, where their availability can limit production, and they can be almost totally depleted in surface waters. Consumption and decomposition of organic matter sinking from surface waters return the nutrients to solution (a process often called *re-mineralization*). As a result, typical concentration–depth profiles look like Figure 2.9.

QUESTION 2.4 (a) Can you suggest why well-stratified surface waters are likely to be more rapidly depleted in nutrients than a well-mixed water column?

(b) Why do the profiles for nitrate and phosphate reach maxima at about 1 km depth, while the maximum for silica is reached somewhat deeper?

All three profiles show some decrease in concentration below the maxima at mid-depth. One explanation of this decline is slow upward mixing from the deep water masses formed at the surface in polar regions, which are themselves relatively depleted in nutrients as a consequence of the biological production occurring there.

Many marine plants and animals form skeletons of calcium carbonate (Section 1.1.1), so carbon is used for both the soft and hard parts of organisms. The biological utilization of carbon and calcium in the marine environment is a major component in the global cycles of these two elements. The amounts used by organisms, however, are small in relation to their total abundance in seawater, and concentration–depth profiles for these two elements—especially calcium—show little depletion in surface waters.

Minor and trace constituents of seawater (hereafter called trace elements for convenience) are also utilized by marine organisms in one way or another. Comparison of Tables 2.2 and 2.1 demonstrates that both major and trace elements can be concentrated to considerable levels in marine organic matter.

QUESTION 2.5 Concentrations in Tables 2.1 and 2.2 are in parts per million. About how many times more concentrated in plankton than in seawater are (a) iron, (b) lead?

Table 2.2 The chemical composition of plankton (in micrograms (10^{-6}g) of element per gram of dry weight of plankton).*

Element	Phytoplankton	Zooplankton
Silicon	58000	—
Sodium	110000	68000
Potassium	12000	11000
Magnesium	14000	8500
Calcium	6100	15000
Strontium	320	440
Barium	110	25
Aluminium	200	23
Iron	650	96
Manganese	9	4
Titanium	≤30	—
Chromium	≤4	—
Copper	8.5	14
Nickel	4	6
Zinc	54	120
Silver	0.4	0.1
Cadmium	2	2
Lead	8	2
Mercury	0.2	0.1

*This Table contains average values, and you may find different values elsewhere, because other authorities use different sources to compile their averages.

The ability to concentrate a trace element from seawater may be shared by a whole class of organisms, but very often uptake is greatest in a particular family or even a single species. Lower organisms usually concentrate trace elements more strongly than higher organisms, but the enrichment mechanisms are complex. Food-chain enhancement, whereby successively higher trophic levels have increased concentrations (as occurs with such environmentally detrimental effects in the case of DDT for example), only happens if the element is actually incorporated into body tissues. Elevated levels of trace elements resulting from concentration within the gut would generally not be passed on to higher trophic levels.

Trace elements are used by various organisms in different ways. For example, copper and vanadium are incorporated in the blood pigments of molluscs and ascidians (sea-squirts); some sponges accumulate titanium; cobalt is used to synthesize vitamin B_{12}; zinc is a constituent of some enzymes; and iron is an important constituent of the blood of many animals (in haemoglobin) as well as being an essential element for plants.

The case of germanium is interesting, because it emphasizes also the importance of **speciation** in the biological reactivity of trace elements. Inorganic forms of germanium behave just like silicon and are incorporated into siliceous skeletons, so the concentration–depth profiles of Si and Ge closely resemble one another. By contrast, organic (methylated) forms of germanium are biologically inert and behave conservatively in seawater. On the other hand, both lead and mercury can be taken up by marine organisms (Table 2.2), but whereas inorganic forms are harmless, the organic forms (alkyl lead, methyl mercury) can be toxic. Speciation is also important in the case of elements that have more than one valency state. Such elements tend to adopt higher valency (oxidized) states in seawater, which contains dissolved oxygen and is therefore an oxidizing medium. When taken up by marine organisms, they are reduced to lower oxidation states (e.g. Mn(IV) to Mn(II), As(V) to As(III), Cr(VI) to Cr(III), Se(VI) to Se(IV)). Upon decomposition of the organic matter, they are released back into solution in the reduced forms, and may be only slowly re-oxidized.

A change in the valency state of an element upon oxidation or reduction can also affect its solubility, and so the **redox** state of the seawater will control the solubility equilibria of such elements. For example, Fe(III) is much less soluble than Fe(II), so in normal seawater the amount of iron in solution is small because most of the iron will exist in particulate form as $Fe(OH)_3$ or FeOOH (much of it small enough to be in colloidal suspension). Iron will become more soluble as conditions become more reducing. Cobalt and manganese are also more soluble in their reduced forms (Co(II) and Mn(II)) than in their oxidized forms (Co(III) and Mn(IV)), which precipitate as oxyhydroxides. Such precipitated phases form part of the inorganic component of particulate matter in seawater. They may contain more than one trace element as a result of co-precipitation (e.g. (Fe,Co)OOH; or $(Mn,Pb)O_2$).

Trace elements incorporated either into the organic matter of soft tissues or into skeletal material will also be more or less depleted in surface waters and enriched in the deep ocean. They are classified as **recycled elements**, most have residence times of less than 10^6 years, and the form of their concentration–depth profiles will lie between two extremes: those of Figure 2.9 at one end, and the near-vertical profile of calcium at the other.

QUESTION 2.6 (a) Sketch the general appearance of the concentration–depth profile for (i) inorganic germanium, and (ii) methylated germanium.

(b) The distribution of cadmium has been found to be well correlated with that of phosphate in the oceans. What might its concentration–depth profile look like? You are *not* expected to put scales on your sketches.

Dissolved constituents in this group can be classified as **biolimiting** or **bio-intermediate**, according to whether or not their availability in surface waters can limit primary production. The nutrients represented in Figure 2.9 are thus biolimiting, while carbon, calcium and barium (Figure 2.1(c)) are bio-intermediate. Convenient though this classification is, however, it should be treated with caution, because it carries the implication that production will cease only when constituents defined as biolimiting are exhausted. This begs the question of the number of potentially biolimiting constituents. For example, there is evidence from work in Antarctic and sub-Arctic waters that production can cease when surface waters become depleted in iron, even though supplies of nitrate and phosphate are still available. Iron seems to be a biolimiting element in this case, and the inference is that in different circumstances there may well be others.

The biolimiting/bio-intermediate distinction is somewhat artificial in another sense, namely that the form of the concentration–depth profile depends not on the absolute amounts used by organisms but on the amounts used *relative to the total quantity available in seawater*. It is clear, for instance, that much more carbon than phosphate is used in biological production; but there is also vastly more carbon in solution and so its concentration–depth profile is all but vertical—yet carbon is no less essential to life than phosphate.

The carbon dissolved in seawater is nearly all inorganic, mainly as the bicarbonate* ion (HCO_3^-), which is the main dissociation product of carbonic acid, H_2CO_3. Dissolved carbon is supplied to the ocean by rivers (mainly as dissolved bicarbonate), by direct solution of CO_2 gas from the atmosphere (forming carbonic acid which then dissociates), and by the respiration of marine organisms, which releases CO_2 into solution, again forming carbonic acid and bicarbonate. We shall look at the carbonate system again in Chapter 3. The important point to recognize here is that the figure for carbon in Table 2.1 represents inorganic forms. The amount of carbon in the dissolved organic matter discussed at the end of Section 2.2.1 is very small by comparison.

Redfield ratios

The average molar ratio of carbon to the two principal nutrient elements in organic matter—nitrogen and phosphorus—is close to 105:15:1 (C:N:P), and this is the basic **Redfield ratio**. The consumption, decomposition and recycling of organic particles as they sink through the water column results in progressive extraction (re-mineralization) of the nutrients and an increase in carbon:nutrient ratios. The particles are said to become more *refractory*. Particles with high settling velocities have a much greater chance of reaching the sea-bed without significant decomposition, and C:N ratios of 6:1 or 7:1 (i.e. with virtually unchanged Redfield ratios) are quite commonly found in the coarser fractions recovered in sediment traps on the sea-bed. Once it reaches the sea-bed, this comparatively fresh material is rapidly decomposed by benthic animals and bacteria.

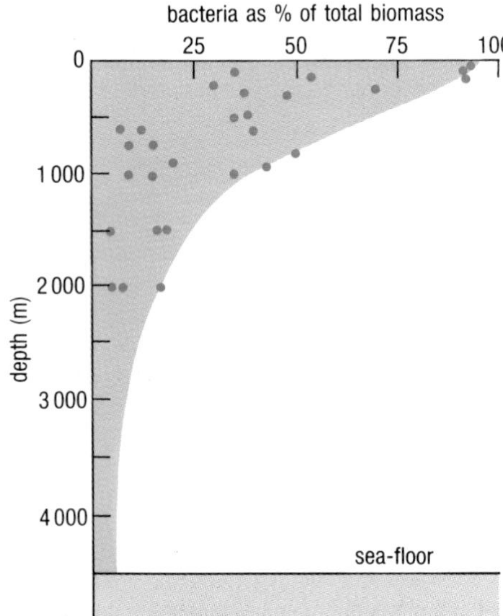

Figure 2.10 Estimates of the percentage (by weight) of the biomass of rapidly sinking organic particles that is contributed by bacteria. Edge of green area represents inferred upper limits. Below about 1–2km depth, most bacteria are probably in free suspension.

We have seen in Section 2.2 that the amount of particulate organic matter decreases with depth, and this must include the total population of bacteria. However, there is some evidence (Figure 2.10) that the *proportion* of bacteria in free suspension, relative to those colonizing larger particles, may actually increase with depth. Below about 2km, bacteria contribute very little to the total biomass of rapidly sinking particles; observation and experiment suggest that such particles may be rather poor habitats for bacteria. This would also help to explain why a significant fraction of such particles reaches the sediments with the nutrient 'load' more or less intact.

Redfield ratios can also be worked out for the main skeleton-forming elements, calcium and silicon (in particulate matter dominated by siliceous organisms, for instance, the ratio is about 105:40:15:1, C:Si:N:P), and for any other elements that are in this recycled group. Deep-water concentrations of many trace elements are strongly correlated with those of the main components in particulate organic matter (*cf.* Question 2.6).

*The term bicarbonate is being replaced by hydrogen carbonate nowadays, but we shall continue to use the shorter term, which is still widespread in the literature.

2.2.3 PARTICULATE ORGANIC MATTER AND THE SCAVENGING CYCLE

The **adsorption** of metal ions or ionic complexes (whether in true solution or in colloidal suspension, e.g. oxides of iron or manganese) onto particle surfaces is a potent mechanism for removing trace elements from the seawater solution. Adsorption results from the mutual attraction between the charges on the ions and suitable bonding sites on the surfaces of the particles. Adsorption is mostly onto the surfaces of bacteria, which make up nearly half of the total amount of particulate organic matter. They are thus very abundant, and their small size ensures that their total surface area is enormous. This is because the ratio of surface area to volume increases progressively with decreasing size: the ratio is about 3×10^3 for a particle of 1mm radius, but 3×10^6 for a particle of 1μm radius. For particles less than 5μm in diameter, the total surface area amounts to about $10 m^2$ per gramme. It has been variously estimated that in the upper ocean something like 90% of the active surface area of suspended particles is on living bacteria; and that the total surface area of freely suspended bacteria is two orders of magnitude greater than that of inorganic particles.

Elements adsorbed onto small particles are removed from the water column as large particles capture the small ones and carry them downwards—a process called **scavenging** (many oceanographers use this term to encompass both parts of the process: adsorption and capture). It is important to stress that adsorption and scavenging are passive processes so far as the bacteria are concerned. They act merely as an agent of transport, they do not make use of the elements they adsorb from solution. Nor are the trace elements adsorbed onto particle surfaces generally in great demand for the metabolic requirements of other marine organisms; and though they will be ingested along with the particles by filter-feeding organisms, they are soon excreted again, to become available for further scavenging.

How efficient is adsorption and scavenging as a way of removing trace elements from the water column?

It is extremely efficient. The mean lifetime of small particles in seawater is about 7.5 years (Section 2.2.1). The mean residence times for **scavenged elements** are rather longer, because adsorption is not a one-way process: there are adsorption–desorption equilibria between ions in solution and those attached to particles; so scavenging cannot be 100% efficient. All the same, the residence time for the scavenged elements is invariably less than 10^3 years, and for many it is less than 100 years.

In some cases, removal from the water column can be extremely rapid. Following the Chernobyl nuclear incident in 1986, sediment traps in a number of places round Europe recorded dramatic increases in concentrations of radionuclides arriving at the sea-bed, within only weeks of the explosion. As soon as the fall-out reached the sea, surface-active nuclides (such as ^{137}Cs, ^{95}Nb, ^{95}Zr, ^{144}Ce, ^{103}Ru) were adsorbed onto particles which were in turn packaged into faecal pellets that sank rapidly to the sea-bed, perhaps more rapidly than would normally have been the case, because the accident more or less coincided with the spring plankton blooms.

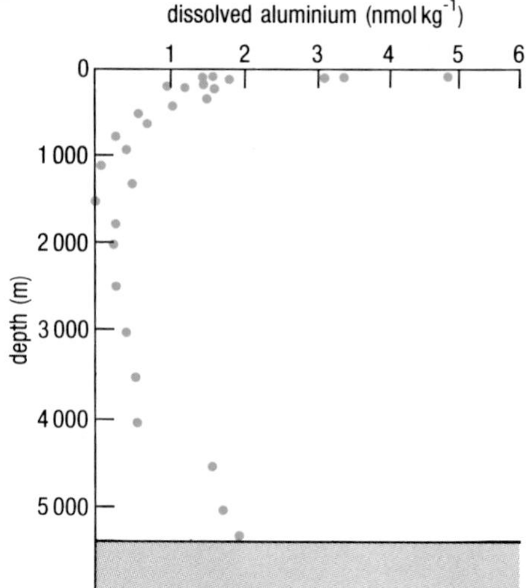

Figure 2.11 Concentration–depth profile for dissolved aluminium in the central North Pacific (28°15′N, 155°07′W). The increase in concentration at the bottom of the profile may be due to re-solution in deep water and/or to diffusion from sediment pore waters (see Chapter 5). (nmol = nanomol = 10^{-9} mole, and nmol l^{-1} ≈ nmol kg^{-1}.)

What does all this tell us about the extent to which scavenged elements are mixed in the oceans?

As the mean oceanic mixing time is about 500 years, and residence times of most scavenged elements are much less than this, these elements cannot be uniformly mixed throughout the oceans (*cf*. Question 2.2). It follows that their distribution must to a very great extent reflect the influence of their sources either at boundaries or within the water column.

For example, there is more dissolved aluminium in Atlantic than in Pacific surface waters. This is partly because more river water enters the Atlantic relative to its size (the ratio of ocean area to land area drained by rivers entering the ocean is about six times smaller for the Atlantic than for the Pacific); and partly because a major source is aeolian dust, especially in low latitudes, and the Atlantic is well supplied with dust from the Sahara, carried on the North-East Trade Winds. The aluminium is in clay minerals and particulate hydrated oxides ($Al_2O_3.nH_2O$). Some of it goes into solution, but in surface waters of the oceans it is soon absorbed and scavenged by sinking particles, so that its concentration decreases with depth (Figure 2.11).

The behaviour of manganese is generally similar to that of aluminium, but is complicated by the fact that manganese is more soluble as Mn(II) than as Mn(IV), and it is not always easy to distinguish between dissolved and particulate forms (*cf*. Note 5 to Table 2.1). A good deal of manganese is supplied to the oceans by rivers, so concentrations in surface waters tend to decrease away from land. In the open oceans, aeolian dust supplies manganese to surface waters, and in some regions, there is another source of manganese, as you will see in Question 2.7.

QUESTION 2.7 Figure 2.12 shows profiles for dissolved manganese at two different stations in the Atlantic Ocean.

(a) What is the likely principal source of manganese in the region represented by each profile?

(b) Having regard to the location of each profile and the concentration scales, which source is the more important in quantitative terms?

The 'peak' in Figure 2.12(b) does not show up in (a), either because advection of hydrothermal plume waters away from the ridge is in the wrong direction; or because the manganese is scavenged from the plume before it is advected into the area of profile (a); or both. In fact, anomalous concentrations of manganese due to hydrothermal inputs can be detected hundreds of km from the source, before they are eliminated by the combined effects of mixing and particle (bacterial) scavenging.

The form of the profile in Figure 2.12(a) needs further explanation, because at first sight it seems to be a paradox. If aeolian dust is the main source of manganese to the surface ocean, then the manganese must be in particulate form, i.e. as oxidized and insoluble Mn(IV). So how can concentrations of *dissolved* manganese be highest at the surface? There is evidence that manganese dissolves quite readily from dust in surface seawater, and it may be that complex photochemical reactions reduce insoluble Mn(IV) to soluble Mn(II). By contrast, manganese is supplied to the deep oceans by hydrothermal solutions (Figure 2.12(b)) mainly as the more soluble Mn(II). In both cases, the manganese is removed from

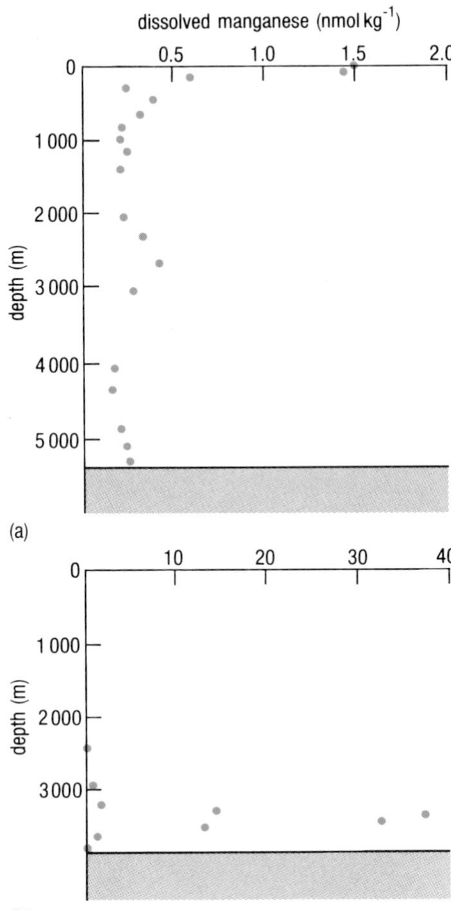

Figure 2.12 Concentration–depth profiles for dissolved manganese, for use with Question 2.7. Both are from stations in the North Atlantic.

(a) 19°N, 23°W.

(b) Above the mid-Atlantic ridge at 26°N.

Note: Each profile has a different horizontal scale.

Table 2.3 Average concentrations (in p.p.m.) of some elements in pelagic clay.

Element	Pelagic clay
Iron	65000
Manganese	6700
Nickel	225
Cobalt	74
Copper	250
Chromium	90
Thorium	12

solution in two ways: partly by direct adsorption and scavenging as Mn(II), followed by re-oxidation to Mn(IV) on particle surfaces; and partly by direct re-oxidation to particulate Mn(IV) in colloidal form, which is then scavenged by larger particles. In either case, it can only be reduced to soluble Mn(II) again if the particles encounter an oxygen-deficient environment (*cf*. Section 2.5.3).

Because the distributions of scavenged elements reflect the influence of sources, they can be used as tracers to help map the movements of **water masses** through the oceans. Although it is a passive process, a great deal of scavenging is biologically driven, particularly where inputs are to surface waters. Good correlations have been found between levels of primary production in the photic zone, the downward flux of particulate organic matter, and the removal of scavenged elements from solution. The Chernobyl fall-out example cited earlier is perhaps a case in point.

2.2.4 A CLASSIFICATION OF THE ELEMENTS IN SEAWATER

Figure 2.13 summarizes in diagrammatic form the characteristic concentration–depth profiles for the three main groups of elements we have just described and discussed, and lists elements representing each group. The conservative constituents (sometimes called accumulated elements) interact only weakly with the biological particle cycle and nearly all have residence times of 10^6 years or more. They include the major constituents discussed in Section 2.1 as well as several trace elements and are sometimes referred to as **bio-unlimited** constituents. The recycled and scavenged groups have been described and discussed in Sections 2.2.2 and 2.2.3 respectively. Moreover, data for the different elements in these groups in Figure 2.13 (overleaf) show that some fit the patterns better than others, and new information could change the position of such elements in the classification scheme.

It is important also to stress that just because an element exhibits a typical recycled profile, it does not mean that the element cannot be scavenged from solution as well (e.g. Ni, V, Cu, Zn, Fe); conversely, a scavenged profile does not preclude involvement in the metabolism of organisms (e.g. Mn, Co, and possibly Al)—concentration–depth profiles merely indicate which of the two patterns of behaviour is dominant. The enrichment of several elements in pelagic clays (Table 2.3) can be attributed in large measure to the effects of particle scavenging in the water column.

It may not be easy to see why the elements fall into the different categories of Figure 2.13, but the distribution is not random. For example, the eleven major constituents of seawater (Section 2.1) all lie in the upper half of the Periodic Table, and eight of them are in Groups I and II (cations) or Groups VI and VII (anions) of the Table; so are hydrogen and oxygen, the components of water itself.

Another interesting regularity concerns what is sometimes called *uptake by analogy*. A number of elements in the recycled group of Figure 2.13 may be there because of their chemical similarity to more abundant vertical neighbours in the Periodic Table, rather than because they have an essential biological role (indeed, some of them may even be toxic). Such pairs include As and P; Ag and Cu; Pd and Ni; Cd and Zn. The last is particularly interesting, because in Section 2.2.2 you read that cadmium

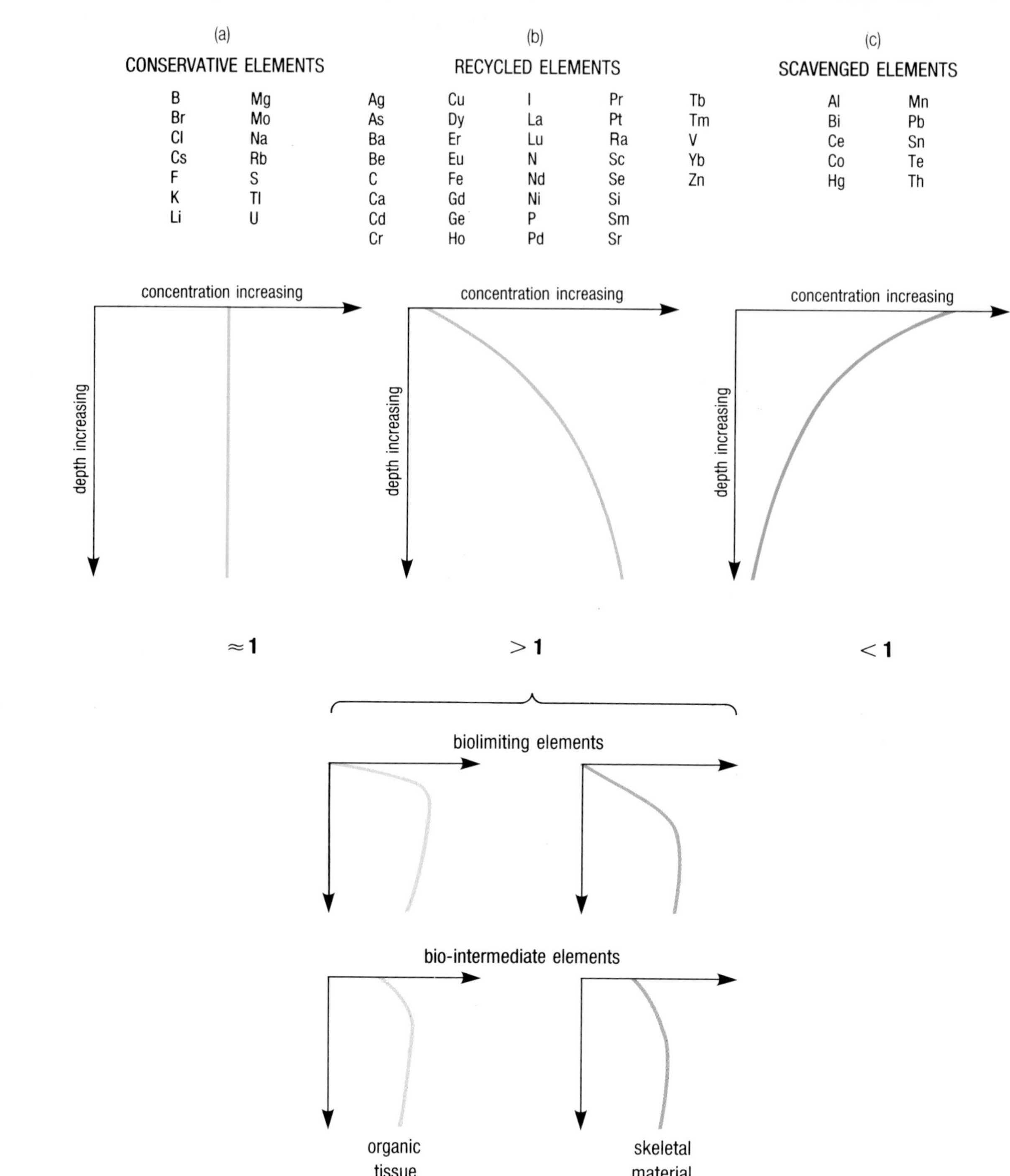

Figure 2.13 Summary classification of elements in seawater according to concentration–depth profiles, with elements listed in each category. For discussion of concentration ratios in deep ocean waters, see Section 2.4.

(a) Conservative (accumulated) or bio-unlimited elements.

(b) Recycled elements. (The recycled group can be further subdivided according to criteria discussed in Section 2.2.2.)

(c) Scavenged elements.

is closely correlated with phosphate in seawater (*cf.* Question 2.6, Figure A1). Although cadmium has no known metabolic role, it may find its way into organisms via the pathway used for zinc and is then 'trapped' there, which would account for its involvement in the cycling of biogenic particulate matter.

QUESTION 2.8 Whereabouts in the Periodic Table would you expect to find (a) strontium in relation to calcium, (b) germanium in relation to silicon?

The scavenging cycle and sediment fluxes

Rates of deposition and accumulation of sediment can be directly measured by deploying sediment traps at various depths down to the sea-bed. Rates of sediment accumulation can be determined by dating successive layers using natural radio-isotopes, such as carbon-14 (^{14}C) and the daughter products of uranium and thorium decay. Sediment fluxes can also be estimated by making use of the scavenging cycle. The principles are simple, but the procedures and calculations are rather complicated, so we shall look briefly at only a couple of examples. Before we begin, it is important to note that concentrations of radio-isotopes are generally measured in physical rather than chemical units: disintegrations per unit time per unit mass.

Uranium behaves conservatively in seawater; isotopes of thorium are among its decay products, and thorium falls in the scavenged category (Figure 2.13). One of the parent–daughter isotope pairs in this series is ^{238}U–^{234}Th. By measuring the concentration of uranium in the water column at a particular depth, the expected concentration of ^{234}Th can be calculated. The actual concentration of ^{234}Th will be less than the calculated value by an amount that depends on how much has been scavenged from solution, and this can in turn be related to the vertical sediment flux at that depth.

Another parent–daughter pair is ^{226}Ra–^{210}Pb. The distribution of ^{210}Pb is complicated by the fact that it is also produced in the atmosphere via the decay of radon gas which escapes from rocks and soils; so, seawater also contains a contribution from atmospheric fall-out, amounts of which seem now to be fairly well known. Lead is one of the scavenged elements in seawater (Figure 2.13), so ^{210}Pb is rapidly removed to sediments by particulate organic matter. The greater the sediment flux, the greater the proportion of ^{210}Pb scavenged from the water column and the greater the flux of ^{210}Pb to the sediments, relative to its rate of supply from the decay of ^{226}Ra and from the atmosphere (Figure 2.14).

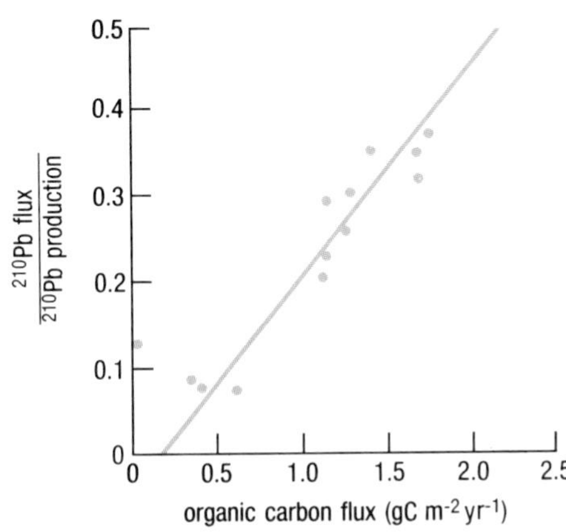

Figure 2.14 Average annual organic carbon flux correlates strongly with the ratio: ^{210}Pb flux into the sediment traps, divided by annual production (including atmospheric deposition) of ^{210}Pb in the water column above the traps. Data are for traps deeper than 3000m.

An advantage of this method is that ^{210}Pb behaves conservatively once it reaches the sediments, i.e. it is buried without participating to any measurable extent in interactions between seawater and sediment pore waters. Some ^{210}Pb is produced by decay of ^{226}Ra in the bottom water and within the sediment, and this contribution can be determined from measurements of the ^{226}Ra concentration. Any excess ^{210}Pb in the sediments must have come from scavenging by particulate organic matter in the water column. The greater this excess, the greater the scavenging, and the greater the sediment flux. The half-life of ^{210}Pb is about 22 years, which means that amounts of this isotope in the topmost layer of a sediment core can provide information about sediment fluxes over the last 100 years or so. This is important for comparing present-day fluxes of carbon to the deep sea with those of the immediate past; and it can be particularly useful where most of the organic matter reaching the sediments has been consumed by animals and bacteria.

Another way of estimating sediment fluxes is to try to quantify the movements of dissolved constituents as they travel through the oceans from sources to sinks. This is the subject of the next Section.

2.3 VERTICAL MOVEMENT OF DISSOLVED CONSTITUENTS: THE TWO-BOX MODEL

Marine chemists find it convenient to treat the oceans as a layered series of well-mixed reservoirs. Here we shall confine ourselves to only two such reservoirs, but this will nonetheless enable us to quantify the basic processes occurring in the oceans and to estimate, to a first approximation, the rates at which different constituents move through the system. We can simplify the oceans as a whole into two reservoirs, a thin upper layer of warm water, with a much larger cold reservoir beneath it.

What is the boundary separating these two reservoirs?

It is the base of the **mixed surface layer**, between 100 and 200m depth on average. This is the bottom of the upper 'box'. Below it lies the permanent thermocline extending down to 500–1000m over most of the world's oceans, and below that are the intermediate and deep water masses. These all make up the lower 'box' which is thus some twenty times bigger than the upper 'box'. This simple **two-box model** can be more easily used if we make further simplifying assumptions:

1 Dissolved constituents are added to seawater by rivers only. Sources such as aeolian, volcanic or hydrothermal inputs, or the inflow of groundwater along continental margins, are ignored.

2 The only way dissolved constituents are removed from the ocean is by organic (biogenic) particles falling to the sea-floor.

3 The ocean is in a steady state: rates of input and loss of any dissolved constituents, both in the oceans as a whole and between the warm and cold reservoirs, have remained constant for long periods, so that concentrations at any point do not change with time.

It follows from this set of assumptions that dissolved material *added* by rivers (the source) to the sea must be *removed* at the same rate, by preservation in sediments accumulating on the bottom (the sink), having been mixed and cycled within the oceans.

QUESTION 2.9 It also follows from those assumptions that the two-box model can be applied only to dissolved constituents in the recycled category of Figure 2.13, but not to the others. Why is that?

We shall now set up and explain the two-box model in diagrammatic form, and then use it to examine the behaviour of a biolimiting element.

Study Figure 2.15 and read its caption carefully, to make sure that you understand the balance between the different fluxes. After reading the caption, you may well have some questions that require answering, but we hope to cover these when we apply the model quantitatively in the following exercise.

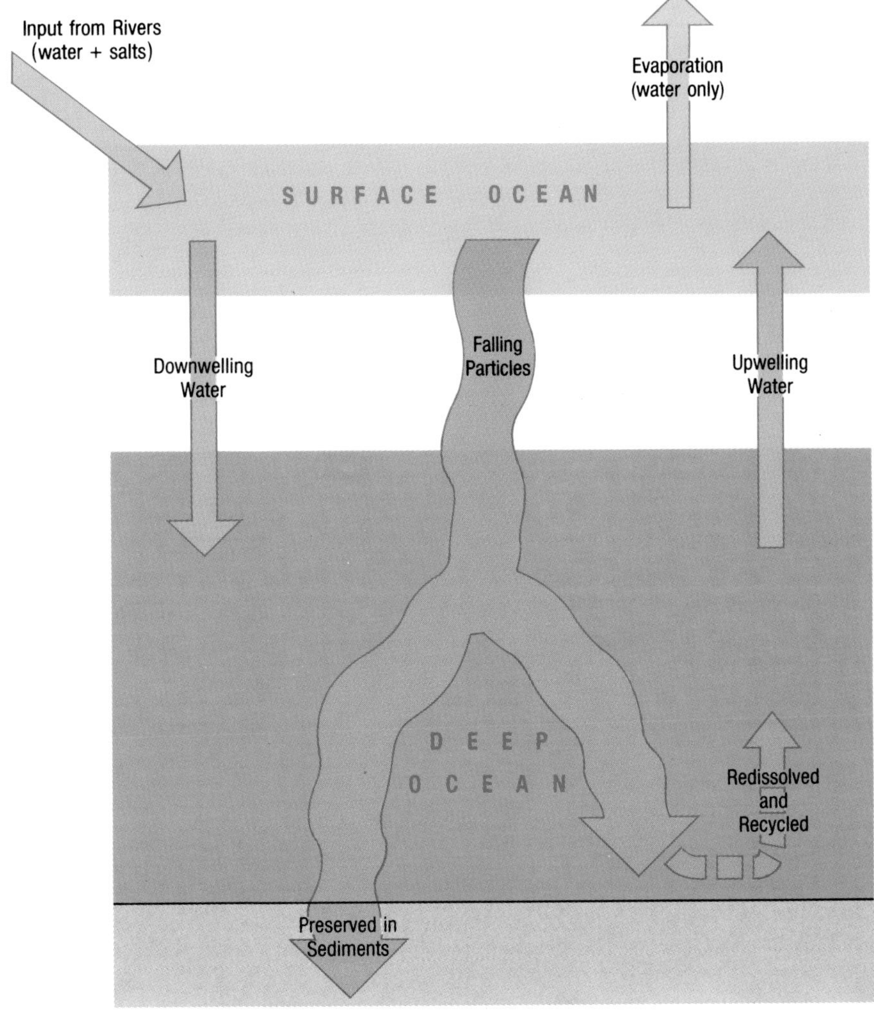

Figure 2.15 Simple two-box model for the oceans. The following notes explain the various fluxes and their relationships:

1 The net Input from Rivers consists of dissolved constituents only, because the hydrological cycle ensures that an equal amount of water is removed by Evaporation.

2 The material supplied by rivers in solution represents new material that must be removed from the system if a steady state is to be maintained. So the Input from Rivers is balanced by material Preserved in Sediments.

3 Material from the surface ocean is lost to the deep ocean in Downwelling Water (partly simple downward mixing, but mainly the formation of deep water masses) and in Falling Particles.

4 The volume of water in both reservoirs must remain unchanged, however, so the Downwelling Water must be balanced by an equal amount of Upwelling Water (this is mainly simple upward mixing, but includes localized upwelling).

5 When the Falling Particles reach the bottom of the deep reservoir, a proportion equivalent to the Input from Rivers is Preserved in Sediments, but the remainder will be Redissolved and Recycled, to be carried up to the surface ocean in due course by the Upwelling Water.

6 *Very important*: As biologically active elements have different concentrations in surface and deep oceans (e.g. Figure 2.9), then very different amounts of these elements will be transported by Downwelling and Upwelling Waters.

2.3.1 THE TWO-BOX MODEL AND PHOSPHATE

Figure 2.16 is a larger version of Figure 2.15, and it contains spaces for you to insert the values of the fluxes and balances, after following through the simple arithmetical calculations set out below. The completed version of Figure 2.16 is provided at the back of this Volume (Figure A2).

Throughout this exercise, we shall use molar concentrations. Because 1 litre of seawater weighs nearly 1 kilogramme, concentrations in $\text{mol}\,l^{-1}$ are virtually the same as concentrations in $\text{mol}\,kg^{-1}$.

Figure 2.16 Two-box model for use with the phosphate exercise. For the completed calculation, see Figure A2, immediately before the answer to Question 2.10.

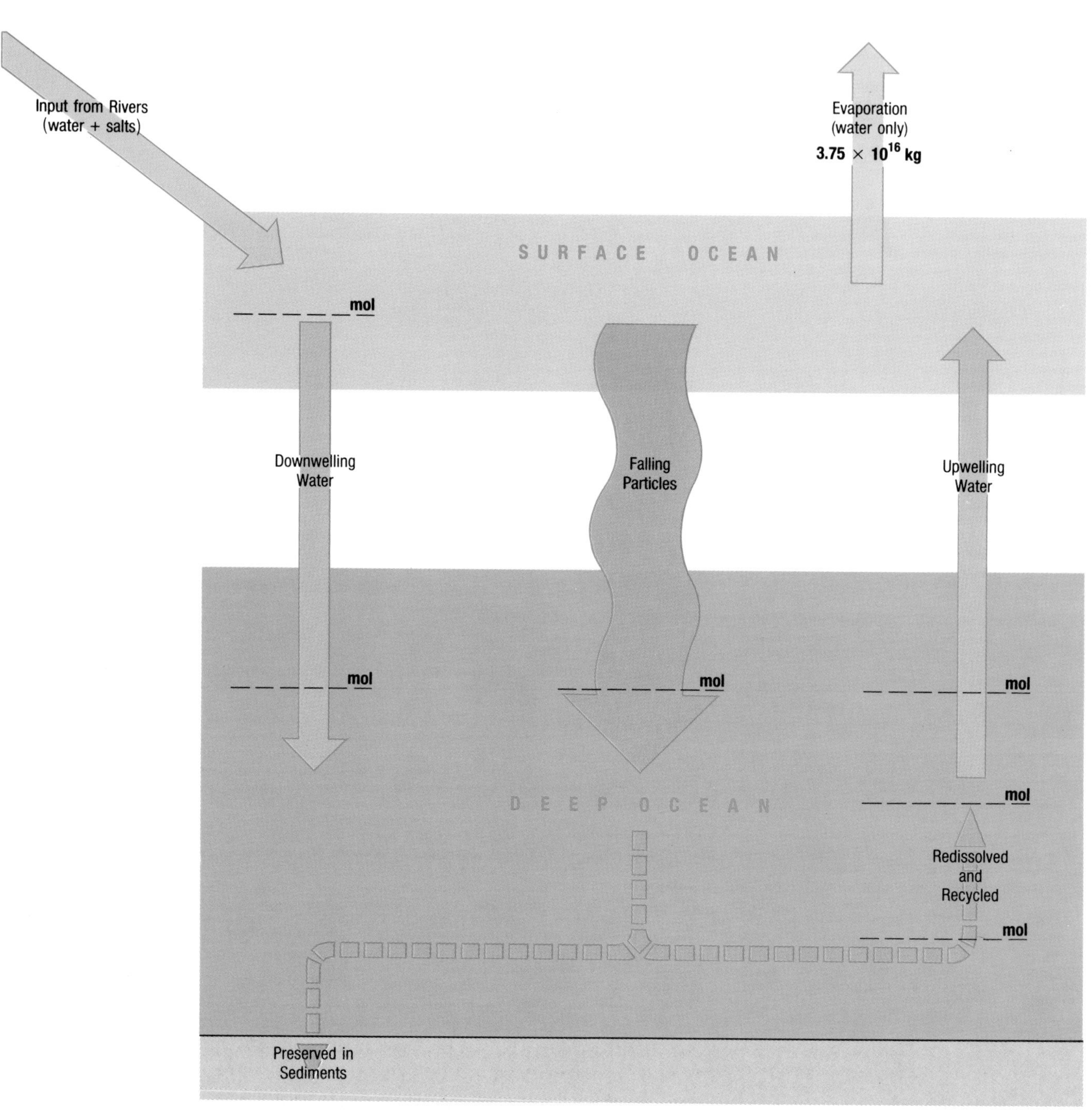

To calculate the Input from Rivers, we need to know the annual flux of water from all the world's rivers into the oceans, and the concentration of phosphate in river water. Some 3.75×10^{16} kg of water are supplied annually to the oceans by rivers. The average concentration of phosphate in river water is about 0.5×10^{-6} mol kg^{-1}, so we can work out the annual Input from Rivers (in moles) and insert the result in Figure 2.16:

Input from Rivers = $(3.75 \times 10^{16}\,\text{kg}) \times (0.5 \times 10^{-6}\,\text{mol } PO_4^{3-}\,\text{kg}^{-1}) = 18.75 \times 10^{9}\,\text{mol } PO_4^{3-}$.

(Note that the Evaporation figure is the *same* as the annual river flux, and we have inserted this already.)

The Input from Rivers must be the *same* as the amount Preserved in Sediments, for the steady-state composition of the ocean to be maintained. So, you can make another entry in Figure 2.16: $18.75 \times 10^{9}\,\text{mol } PO_4^{3-}$ Preserved in Sediments.

Oceanographers have used carbon-14 age-dating techniques to work out that the upwelling flux of seawater is about 20 times the river flux. Upwelling and downwelling fluxes must be equal, and so you can easily work out the amount of *water* that is being exchanged between the surface and deep oceans each year.

What is it?

As the river flux is about $3.75 \times 10^{16}\,\text{kg yr}^{-1}$, then both the upwelling and downwelling flux must be in the order of $7.5 \times 10^{17}\,\text{kg yr}^{-1}$.

What is the principal feature of a biolimiting recycled constituent such as phosphate with respect to its concentrations in the surface and deep oceans?

Concentrations are very much less in the surface than the deep oceans, and for phosphate these concentrations are in the order of $0.1 \times 10^{-6}\,\text{mol kg}^{-1}$ and $2.5 \times 10^{-6}\,\text{mol kg}^{-1}$ respectively (Figure 2.9).

You now have the water flux for Downwelling and Upwelling Water, and the phosphate concentrations in the surface and deep oceans, so you can work out respective phosphate fluxes for Figure 2.16:

Downwelling Water = $(7.5 \times 10^{17}\,\text{kg}) \times (0.1 \times 10^{-6}\,\text{mol } PO_4^{3-}\,\text{kg}^{-1}) = 75 \times 10^{9}\,\text{mol } PO_4^{3-}$

Upwelling Water = $(7.5 \times 10^{17}\,\text{kg}) \times (2.5 \times 10^{-6}\,\text{mol } PO_4^{3-}\,\text{kg}^{-1}) = 1875 \times 10^{9}\,\text{mol } PO_4^{3-}$

If the concentration in the surface ocean is to be maintained at its low level, then the huge difference between the phosphate fluxes for Upwelling Water and Downwelling Water must be balanced.

By what?

Most of the phosphate that is added from rivers and that wells up from the deep ocean is fixed by organisms, which subsequently die and sink, carrying the phosphate with them. So they transport the balance of the phosphate that is not carried to the deep ocean by Downwelling Water.

The amount of phosphate carried down by Falling Particles must therefore be given by:

$$PO_4^{3-} \text{ in Particles} = PO_4^{3-} \text{ in Upwelling} + PO_4^{3-} \text{ in Rivers} - PO_4^{3-} \text{ in Downwelling}$$
$$= (1875 \times 10^9) + (18.75 \times 10^9) - (75 \times 10^9)$$
$$= 1818.75 \times 10^9 \text{mol} PO_4^{3-}$$

Now all that remains is to work out the amount of phosphate that is Redissolved and Recycled in the deep ocean.

What will that amount to?

It will be the difference between the amount of phosphate carried down by Falling Particles, and the amount that is Preserved in Sediments:

$$(1818.75 \times 10^9) - (18.75 \times 10^9) = 1800 \times 10^9 \text{mol } PO_4^{3-}$$

As a check: PO_4^{3-} in Upwelling = PO_4^{3-} Redissolved and Recycled + PO_4^{3-} in Downwelling:

$$(1875 \times 10^9) = (1800 \times 10^9) + (75 \times 10^9)$$

(The completed two-box model for phosphate is given in Figure A2, before the answer to Question 2.10.) To make useful comparisons of the results of two-box model calculations for different constituents, we need to work out:

1 What percentage of the constituent entering the surface ocean (a) goes into Falling Particles, (b) is Preserved in Sediments.

2 What percentage of the constituent in Falling Particles is Preserved in Sediments.

QUESTION 2.10 Now do these additional calculations (1 and 2) for phosphate.

So far as phosphate is concerned, then, nearly all of the phosphate that enters the surface ocean is removed in Falling Particles. However, only about 1% of the Falling Particles survives to be Preserved in Sediments—99% is Redissolved and Recycled. Only about 1% of the phosphate entering the surface ocean (from *both* Rivers *and* Upwelling) is Preserved in Sediments. Put another way, each mole of phosphate goes through an average of about 100 cycles within the oceans before it is incorporated into sediments on the ocean floor.

Whereabouts does most of this recycling occur?

Most nutrient recycling occurs in the mixed surface layer, i.e. in the upper box, because primary production can take place only within the photic zone (*cf*. Figure 2.3). Below this, organic matter is consumed and decomposed, returning the nutrients to solution (Figure 2.9).

2.3.2 ESTIMATING RESIDENCE TIMES

One of the essential pieces of information for two-box model calculations is the figure for Input from Rivers. From the definition given earlier:

$$\text{residence time} = \frac{\text{mass of element in oceans}}{\text{rate of input or removal}}$$

We can use the figure for Input from Rivers with data from Table 2.1 to estimate residence times.

Caution: The residence time calculation is simplicity itself, but you must be careful to use the right numbers. In Section 2.3.1, concentrations and quantities are in molar terms, in Table 2.1 they are in terms of weight. The relationship is:

molar quantity $\times$ (relative atomic molecular mass) = weight quantity (in grammes)

The relative mass of phosphorus is 31. On average, 18.75×10^9 mol of PO_4^{3-} enter the oceans from rivers each year (Input from Rivers, Figures 2.16 and A2). This is equivalent to $18.75 \times 10^9 \times 31 \approx 580 \times 10^9$ g of P, which is 580×10^3 tonnes annually (remember that 1 mole of phosphate (PO_4^{3-}) contains one mole of phosphorus (P)). The total mass of phosphorus in the oceans from Table 2.1 is 7.93×10^{10} tonnes, and so the residence time must be:

$$\frac{7.93 \times 10^{10}}{580 \times 10^3} = \frac{793 \times 10^8}{580 \times 10^3} \approx 1.4 \times 10^5\text{, or 140000 years}$$

(assuming that all phosphorus is in the form of phosphate).

Information from two-box model calculations can also be used to estimate residence times. We have seen that each mole of phosphate goes through an average of 100 cycles within the oceans before it is incorporated into sediments on the ocean floor. Each cycle has a duration of 500 years, the average mixing time of the oceans as a whole. On that basis, the average residence time for phosphate is:

$100 \times 500 = 50000$ years.

This is much less than the estimate arrived at by the 'standard' way, but that is hardly surprising in view of the considerable amount of averaging we have used in both estimates (see also Section 2.4).

The two-box model is a greatly simplified approximation of the real ocean. Powerful computers now enable marine scientists to work with vastly greater numbers of well-mixed (and not so well-mixed) reservoirs—by the mid-1980s, an 86-box model had been developed.

Despite the approximations and assumptions of our simple two-box model, however, the *basic principles* are valid, and the results are realistic enough to bring out the more important contrasts in the behaviour of different recycled constituents, *provided always that you must not expect to be able to quote those results to umpteen significant figures.*

2.4 LATERAL VARIATIONS OF DISSOLVED CONSTITUENTS IN THE DEEP OCEANS

One obvious feature of the two-box model which may not have escaped your notice is that it emphasizes vertical fluxes and pays no attention to lateral variations. For example, the two-box model by itself might well give the impression that the small proportion of phosphate leaving the cycle each year to be Preserved in Sediments, is uniformly spread throughout the oceans as a whole. Not so.

Figure 2.17 shows how the concentration of phosphate varies in the deep oceans. From a minimum in the North Atlantic, it increases southwards, and continues to increase round southern Africa, eastwards and north-eastwards in the Indian Ocean, and also eastwards and northwards in the Pacific. Table 2.4 summarizes concentration ratios in the Atlantic and the Pacific for four elements used in biological processes. Note that the ratio uses the *difference* between deep and surface concentrations in each ocean, so Table 2.4 is a measure of the *lateral* enrichment of each element in deep Pacific water compared with deep Atlantic water.

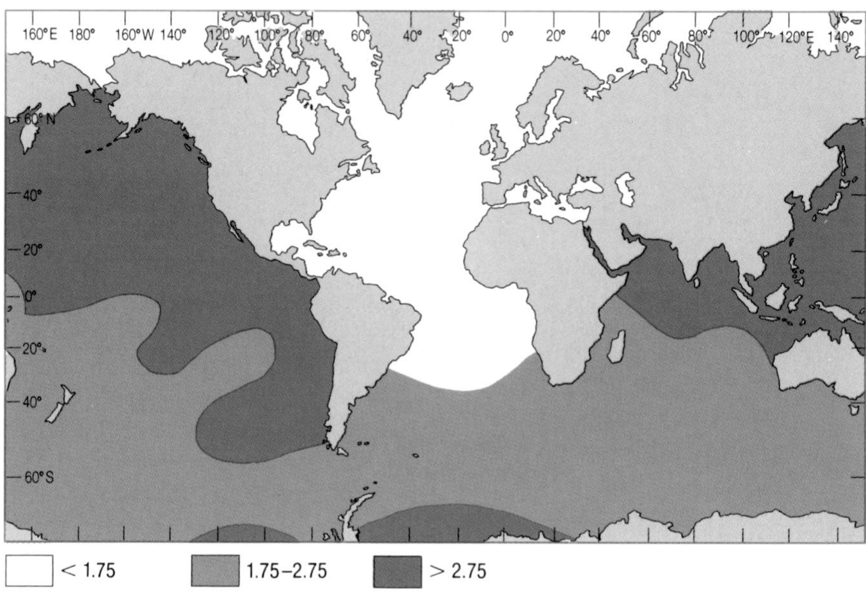

Figure 2.17 Distribution of phosphate in the oceans at 2000m depth.

QUESTION 2.11 Is there any obvious difference between the ratios for hard-part and soft-part constituents in Table 2.4?

Table 2.4 Approximate average ratios for biologically important elements in the deep parts of the Atlantic and the Pacific (C=concentration).*

Element	$\frac{(C_{deep} - C_{surface})\ \text{Pacific}}{(C_{deep} - C_{surface})\ \text{Atlantic}}$
Nitrogen (as NO_3^-)	2
Phosphorus (as PO_4^{3-})	2
Carbon	3
Silicon	5

*Note that the geographic distribution of deep-ocean enrichment for each element is qualitatively similar to that for phosphate (Figure 2.17); and on a global scale surface concentration in each ocean will be broadly similar for each element.

Bearing in mind that the ratios in Table 2.4 express an enrichment in the *deep* oceans (which is also illustrated in Figure 2.17), we need to look for a mechanism to explain why the deep Pacific is so much richer in nutrients than the deep Atlantic.

This mechanism is summarized in Figure 2.18. Nutrient-poor **North Atlantic Deep Water** (NADW) flows southward in the western Atlantic and is steadily enriched with nutrients derived from the rain of particulate organic matter sinking from the surface and being redissolved in the deep ocean. It then flows round southern Africa and into the Indian and Pacific Oceans. In the southern Atlantic it is joined by **Antarctic Bottom Water** (AABW), which is more nutrient-rich than NADW to start with, because of the upwelling along the Antarctic Divergence. Although some AABW flows north in the Atlantic, much of it flows into the Indian and Pacific Oceans, along with the NADW, and nutrient enrichment by sinking particulate organic matter continues. Thus, the oldest and most nutrient-rich waters are in the deep Pacific Ocean (Table 2.4).

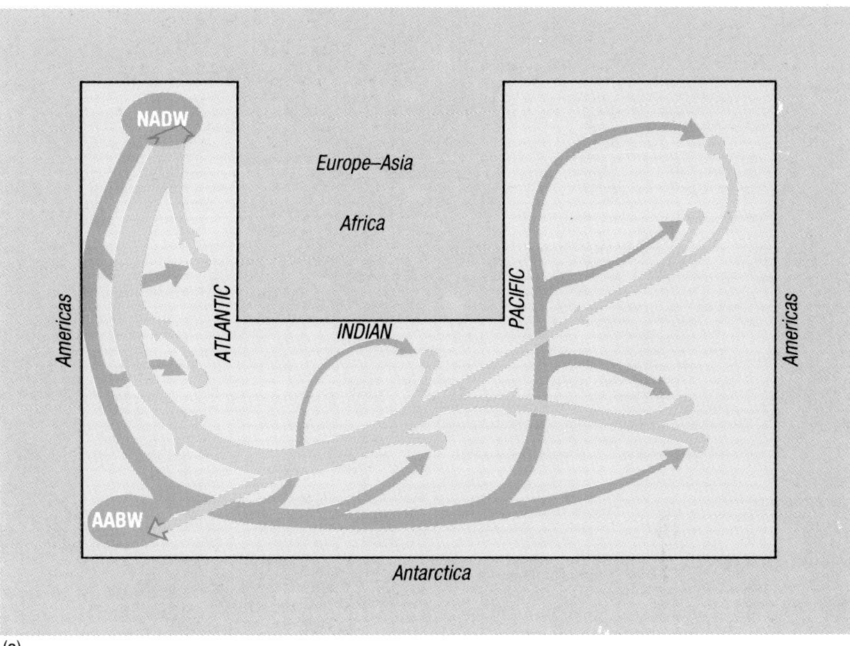

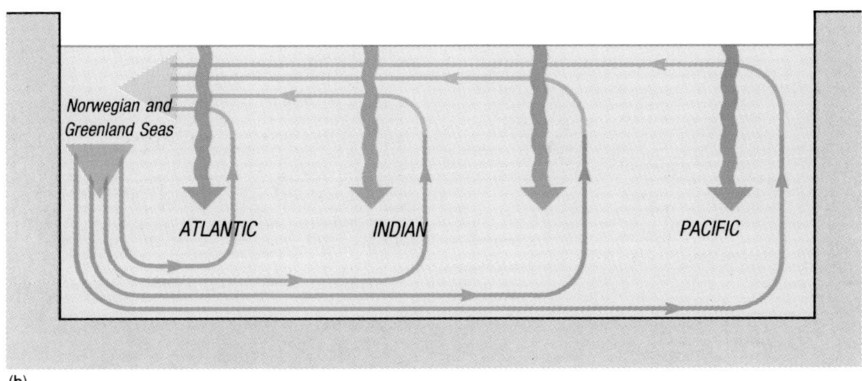

Figure 2.18 (a) Generalized map of deep water flow (*dark blue*) and surface water return (*mid-blue*) in the oceans. Large ellipses designate sources of North Atlantic Deep Water (NADW) and Antarctic Bottom Water (AABW); small mid-blue circles indicate areas of localized upwelling.

(b) Generalized cross-section from the North Atlantic to the North Pacific, showing major advective flow patterns (*thin pale blue lines*) and the rain of particles (*wavy arrows*). AABW is not shown in this picture.

Deep water is returned to the surface mainly by the broad, slow and rather diffuse upwelling that goes on continuously throughout the World ocean. The localized upwelling areas schematically identified in Figure 2.18 are also important, but much of the upwelled water in these regions comes from intermediate depths and is not truly 'deep water'.

All of this clearly has some bearing on the significance of a mean oceanic stirring time, and suggests that the deep water in some parts of the oceans is much older than our average value of about 500 years. Radiocarbon measurements of deep water in parts of the northern Indian and Pacific Oceans yield values in excess of 1000 years, while in other regions the exchange between deep and surface water takes much *less* than 500 years. As is so often the case, the average thus conceals considerable variations, which exemplifies the points made at the end of Section 2.3.2.

QUESTION 2.12 In general, the older the deep water, the richer it will be in elements with nutrient-type or recycled concentration–depth profiles. That explains the concentration ratio for this category in Figure 2.13. Can you explain why the ratio for scavenged elements is less than 1?

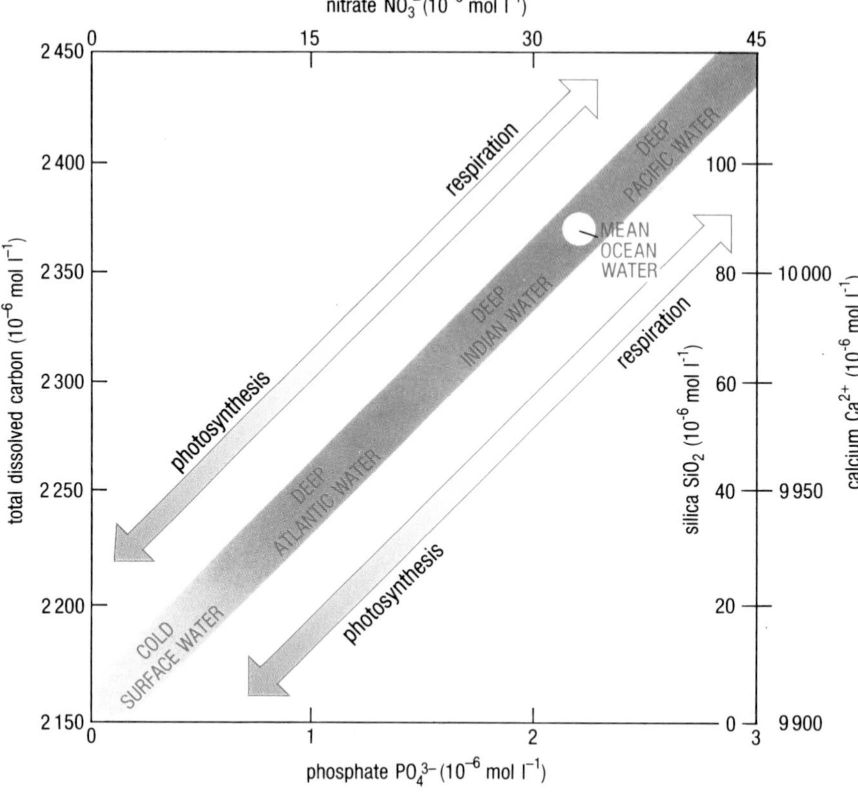

Figure 2.19 Idealized diagram to illustrate changes in nutrient and skeleton-building elements in the oceans as a result of biological activity. Conditions at the surface-water end (*lower left*) of the diagonal line are achieved when the biolimiting nutrients have been exhausted as a result of photosynthetic activity by phytoplankton. The other end of the line is determined by the degree of enrichment in the deep sea from decomposition of organic matter as a consequence of respiration by marine organisms. Intermediate values are produced by mixing these end-members in all proportions. Data for Ca and Si are very generalized and apply only where hard parts are formed. Where there are organisms with calcareous hard parts, proportionately more C would be used in the formation of $CaCO_3$.

Figure 2.19 summarizes the overall distribution of the five biologically most important dissolved constituents of seawater in different parts of the ocean. You should note that the molar ratio of N:P is shown as remaining constant throughout the range of Figure 2.19. The molar N:P ratio of organic tissue is also close to 15:1 (the Redfield ratio, Section 2.2.2). Thus, when all the dissolved nitrate in surface waters has been used up, so has all, or nearly all, the dissolved phosphate—and vice versa. The question of why nitrate and phosphate should occur in seawater in the same ratio that organisms require them remains unanswered. We do not know whether the 15:1 molar ratio of N:P occurred at the beginning of evolution and marine organisms adapted to it, or whether the organisms have themselves established the ratio through time.

2.5 THE AIR–SEA INTERFACE

The transfer of heat and water across the ocean–atmosphere boundary is of great importance in controlling the Earth's climate. Transfers of gases, liquids and solids across the interface are important components of the chemical cycles that occur within the oceans (Figure 2.2). The so-called *stagnant boundary layer* across which these transfers take place is only a few tens of μm thick. Hydrophobic (water-repellent) organic molecules such as fats and oils, both natural and artificial, tend to accumulate at the interface itself, the top of the stagnant boundary layer. The layer itself is almost completely transparent to solar radiation, and just below it, in the uppermost few millimetres of the water column, live the *neuston*, which includes phytoplankton specially adapted to survive and utilize high intensities of ultraviolet light. Zooplankton in the neuston are frequently coloured in shades of blue. The neuston escaped detection for a long time, because of the difficulties of sampling at the air–sea interface.

2.5.1 TRANSFER OF GASES

Gases are transferred across a gas–water interface by molecular diffusion. The process is two-way, even at saturation, when the concentration of gas in the water is in equilibrium with the concentration of the gas in the atmosphere, and gas transfer is equal in both directions. The thickness of the boundary layer is not constant: it decreases with increased turbulence in the water, so gas exchange is more rapid during storms than in calm weather. The solubility of all gases is greater in cold than in warm water: if the temperature of the water rises, then under equilibrium conditions there will be a net flux of gas from sea to air. The solubility of gases in water is also increased by pressure; in addition, solubility varies inversely with salinity—the higher the salinity, the lower the solubility.

The basic equation describing the net flux, F, of a gas across an air–water boundary is:

$$F = K\Delta C \tag{2.1}$$

where K is the transfer velocity, the rate at which transfer occurs, and ΔC is the concentration difference driving the gas flux across the interface. K is usually expressed in $cm\,hr^{-1}$, and if C is in $ml\,l^{-1}$, then F will be in $ml\,cm^{-2}\,hr^{-1}$—i.e. unit volume crossing unit area in unit time.

Would you expect values of K in equation 2.1 ever to reach zero?

The form of equation 2.1 is such that K is in effect a proportionality constant (it is often called the exchange coefficient). At saturation, where ΔC is effectively zero, there will still be exchange of gas through the boundary layer, but this will be the same in both directions. The *net* flux, F, will thus be zero. The only conditions under which K itself might approach zero are where a 'lid', such as a thick oil slick, covers the surface. In normal circumstances, values of K range from a few centimetres to a few tens of centimetres per hour, and increase with increased wind speed.

Gases tend to fall into two categories:

1 Gases of low solubility and low chemical reactivity in water, such as N_2, O_2, Ar and other noble gases.

2 Gases of high solubility and/or high chemical reactivity in water, such as SO_2, NO_2 and NH_3.

Major gases

The four main atmospheric gases are nitrogen, oxygen, argon, and carbon dioxide, in that order. To a first approximation, their concentrations in seawater are in equilibrium with their **partial pressure** (which is the same as their percentage by volume) in the atmosphere.

To which of categories 1 and 2 above do these four gases belong?

The first three are in category 1 (low solubility and reactivity). You might expect CO_2 to be in the second category, because it reacts with water to form carbonic acid (see also Chapter 3). However, it reacts more slowly than SO_2, etc., and it is generally grouped with the gases in category 1.

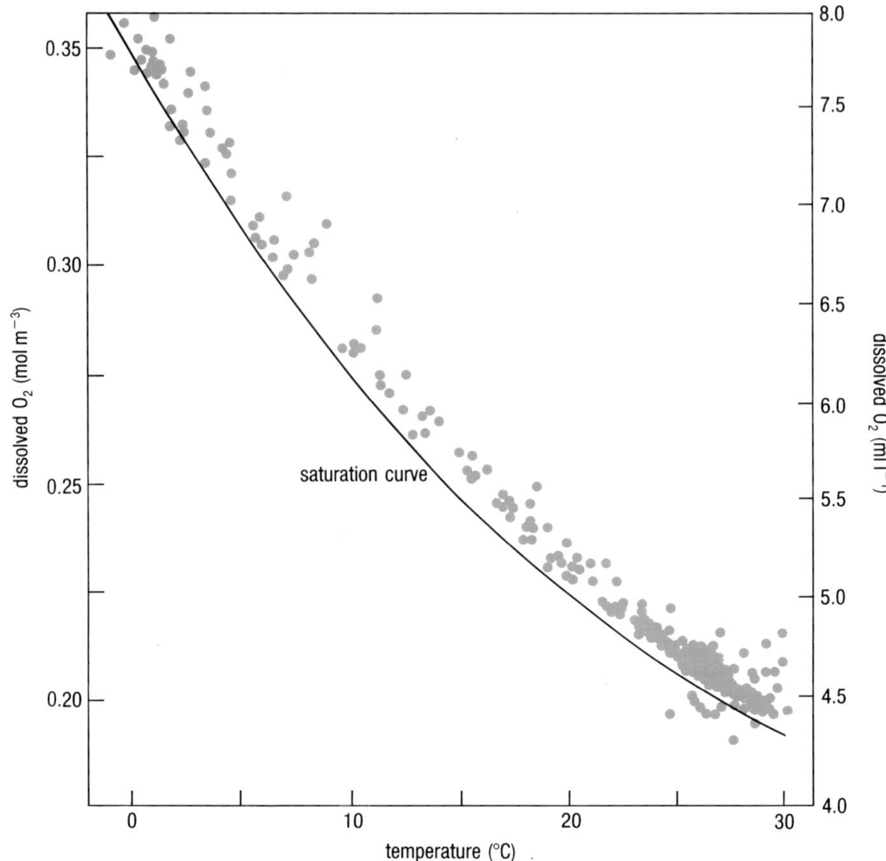

Figure 2.20 The saturation curve for oxygen (black line) and measured concentrations of dissolved oxygen in surface ocean waters (blue dots).

Figure 2.20 shows that oxygen is generally supersaturated in surface seawater. Some of the excess comes from the oxygen released by photosynthesizing phytoplankton. A more important cause of the supersaturation is air bubbles being carried by breaking waves down into the water column, where some of the air dissolves because of the increased hydrostatic pressure. The other atmospheric gases are also slightly supersaturated in surface waters, just as oxygen is.

QUESTION 2.13 Do the four main atmospheric gases behave conservatively or non-conservatively in seawater?

Although carbon dioxide is the most soluble of the major gases, its concentration as dissolved *gas* in seawater is very small. Nearly all the carbon dioxide in seawater is combined with water as carbonic acid and its dissociation products. The solubility relationships of CO_2 are controlled not only by equations such as 2.1, but also by the chemical equilibria governing the reactions of the aqueous carbonate system (see Chapter 3). These reactions can help to 'mop up' excess atmospheric CO_2 produced by human activities and thus act to moderate the global 'greenhouse' warming predicted by many scientists.

Minor gases

These fall into both categories 1 and 2 above: some minor gases behave conservatively, others non-conservatively. It is not easy to quantify transfers of these gases across the air–sea interface, because of the uncertainties inherent in measuring both K and ΔC in equation 2.1. Values of K vary with factors such as temperature and wind speed, and uncertainties in ΔC are due principally to the difficulty of measuring very low concentrations of minor gases in the marine environment (down to 1 part in 10^{12} or less).

For some minor gases, the net flux is from atmosphere to oceans, and this applies particularly to many pollutant gases, which include industrial additions to the natural SO_2 load (*cf.* CO_2 above), and other gases such as CFCs (chlorofluorocarbons) and PCBs (polychlorinated biphenyls).

Gases produced in ocean surface waters by planktonic organisms include carbon monoxide (CO), nitrous oxide (N_2O), and dimethyl sulphide (DMS)—($(CH_3)_2S$)—and for these gases there is a considerable net flux from sea to air.

Why should that be?

They are produced in surface waters and they are supersaturated there. For example, surface waters can have concentrations of DMS as high as $3000 \times 10^{-6} g l^{-1}$ in regions of high biological production; whereas the concentration at equilibrium with normal atmospheric concentrations is three orders of magnitude less (*c.* $3 \times 10^{-9} g l^{-1}$). DMS is oxidized in the atmosphere, forming sulphate aerosols which provide additional nuclei for water droplets and thus aid cloud formation. Some scientists believe that this process can help to counter greenhouse warming: productivity increases as temperature and CO_2 levels rise, more DMS is released, more clouds form, more solar radiation is reflected back to space by the clouds, and temperatures fall. Natural feedback loops of this kind are part of the **Gaia Hypothesis**, whose proponents consider that the surface of the Earth has been maintained as a life-supporting environment for thousands of millions of years largely because of the activities of the plants and animals which inhabit it.

2.5.2 TRANSFER OF LIQUIDS AND SOLIDS

In this context, the liquid is water and it contains both particles and dissolved gases and salts. For air–sea transfer, the flux of material across the interface obviously depends primarily on the intensity of the precipitation (rain, snow or hail), but it is also determined by the **washout ratio**, w, which is defined as:

$$w = \frac{C'}{C} \qquad (2.2)$$

where C' is the concentration of particles or gas in near-surface precipitation, and C is the concentration in the atmosphere.

Washout ratios can be determined experimentally, and they are useful in an environmental context because they are a measure of the ability of water droplets to take up particles or gas. For particles, values of w range from 10^2 to between 10^5 and 10^6. For gases of low solubility and low chemical reactivity in water, w is much smaller, in the range 10^{-2}–1; and for gases that are more soluble and/or chemically reactive, w ranges from 10^2 to 10^4, and exceptionally up to 10^5.

QUESTION 2.14 These values of w have some interesting implications.

(a) Near the sea-surface, are (i) unreactive or (ii) reactive gases more or less concentrated in rain than in the atmosphere?

(b) In general, does rain scavenge particles or gases more efficiently from the atmosphere?

In the absence of rainfall, *dry deposition* of particles to the sea-surface will continue simply by direct fall-out from the atmosphere; the larger the particle, the greater its settling velocity and the more rapid the fall-out. To a very simple first approximation, for particles smaller than about 0.1 μm, such as occur in smoke, the transfer process is essentially diffusive, i.e. the particles behave rather like gas molecules.

The transfer of water and solids from *sea to air* by bubble bursting and aerosol formation should not be forgotten. Aerosol droplets transport not only dissolved salts, but also organic matter from surface layers; some of this is returned directly to the sea by rain or dry deposition, but some must fall on land—so there is a degree of exchange of organic carbon between continents and oceans (*cf.* Section 2.2.1).

2.5.3 OXYGEN IN SEAWATER

Consumption and decomposition of organic matter require oxygen for respiration of the decomposer organisms (animals and bacteria), and this reaches a maximum at around 1 km depth, where there is an **oxygen minimum layer**. The vertical distribution of oxygen in seawater generally approximates to a mirror image of those for the nutrients (Figure 2.21).

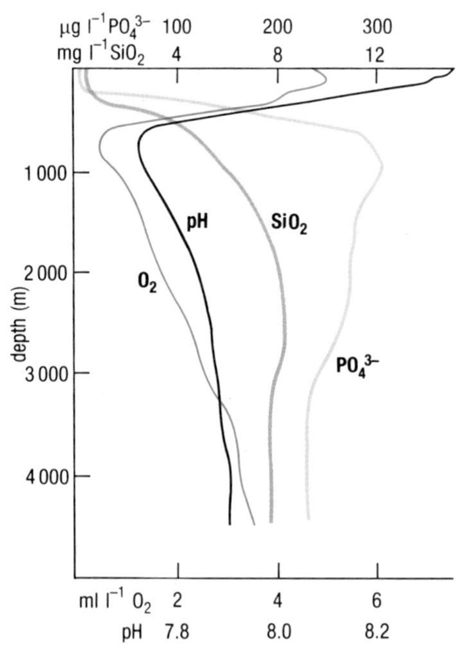

Figure 2.21 Concentration–depth profiles from 24°22′N, 145°33′W, to show the contrasted patterns for nutrients and oxygen. Note that the maximum for silica is reached at greater depth than that for phosphate (*cf.* Figures 2.9 and 2.13). For pH profile, see Question 2.15.

QUESTION 2.15 Can you explain the shape of the profile for pH in Figure 2.21?

We can simplify the essentially reversible reactions that take place during formation of organic matter by photosynthesis, and its subsequent destruction by respiration (oxidation), in the following way:

Representing organic matter by CH_2O (for fixed C) and NH_3 (for fixed N), then:

For carbon:

$$CO_2 + H_2O \rightleftharpoons CH_2O + O_2 \tag{2.3}$$

For fixed nitrogen:

$$NO_3^- + H_2O + H^+ \rightleftharpoons NH_3 + 2O_2 \tag{2.4}$$

These reactions move to the right during photosynthesis, to the left during respiration. From reaction 2.3, it is clear that for every mole of carbon used in photosynthesis, one mole of oxygen is liberated; and for every mole of fixed nitrogen used, two moles of oxygen are liberated.

We can thus include oxygen in the Redfield ratio, which was introduced in Section 2.2.2 and approximates to 105:15:1 (C:N:P).

From reaction 2.3, 105 moles of oxygen (O_2) are liberated (or consumed) for every 105 moles of carbon fixed in (or liberated from) organic tissue. From reaction 2.4, a further 30 moles of oxygen are liberated (or consumed) for the 15 moles of fixed nitrogen that accompany the carbon

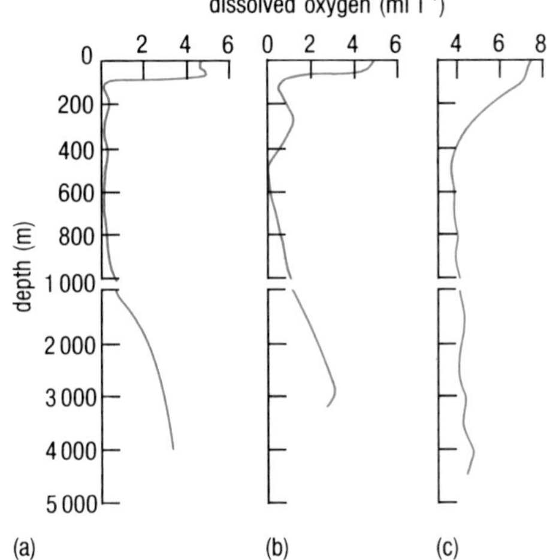

Figure 2.22 Oxygen profiles from three locations, to show the variable form of oxygen minimum layers. *Note the change of scale at 1000m depth.* (a) Eastern tropical Pacific (11°39′N, 114°15′W). (b) Eastern tropical Pacific (6°21′N, 103°42′W). (c) Antarctic Convergence (50°08′S, 35°49′W. Sub-oxic conditions have developed in (a) and (b).

in organic tissue. Phosphorus remains largely as phosphate during biological processes and does not participate in this cycle of reduction and oxidation to any significant extent. We can accordingly extend the Redfield ratio as follows:

105 moles of carbon (C) are equivalent to
30 moles of nitrate (NO_3^-) are equivalent to
1 mole of phosphate (PO_4^{3-}) is equivalent to
} 135 moles of oxygen (O_2)

Armed with this information and a knowledge of how much oxygen and organic matter there is in seawater, we can go some way to explaining semi-quantitatively why oxygen minimum layers can be quite strongly depleted in dissolved oxygen (Figure 2.22).

For example, an oxygen concentration in surface water of 6 ml l^{-1} approximates to 0.27 mol O_2 m^{-3} (Figure 2.20).

What concentration of organic carbon would be required to use up all of that oxygen in respiration (the reverse of reaction 2.3)?

From the Redfield ratio, the amount of organic carbon required to use up this oxygen must be:

$$0.27 \times \frac{105}{135} = 0.21\,\text{mol C m}^{-3}$$

Application of two-box model calculations suggests that the average annual Falling Particle flux of organic carbon amounts to about 0.2 mol C m^{-3}. That is close to the estimate of 0.21 mol C m^{-3} which we have just worked out. So it is not surprising that concentrations of dissolved oxygen can approach zero in oxygen minimum layers (Figure 2.22)—such conditions are sometimes described as *sub-oxic*.

Because gases are more soluble in cold than in warm water, the deep water masses formed at the surface in high latitudes are the richest in dissolved oxygen. As the water masses sink and move away from their source regions, the oxygen is progressively used up by marine organisms in respiration.

QUESTION 2.16 With reference to Figure 2.18, where would you expect to find the lowest concentrations of dissolved oxygen in oceanic bottom waters?

Nonetheless, even in the oldest bottom waters, oxygen concentrations rarely fall much below about 3 ml l^{-1}, so the deep sea-bed is kept well-oxygenated throughout the major ocean basins. The water at mid-depths can become very oxygen-deficient (Figure 2.22), and such water can be extensive in some oceanic areas. However, it is rare for any part of the water column in the open oceans to become completely devoid of oxygen, but this can happen in the shallower waters of continental shelf areas.

Anoxic environments in seawater

Water depths in most continental shelf areas are less than 200 m, which approximates to the average depth of the mixed surface layer. Under normal conditions, therefore, the waters of continental shelves are well-oxygenated because of mixing by waves and tidal currents. However, in regions of high productivity, where the amount of sinking organic matter

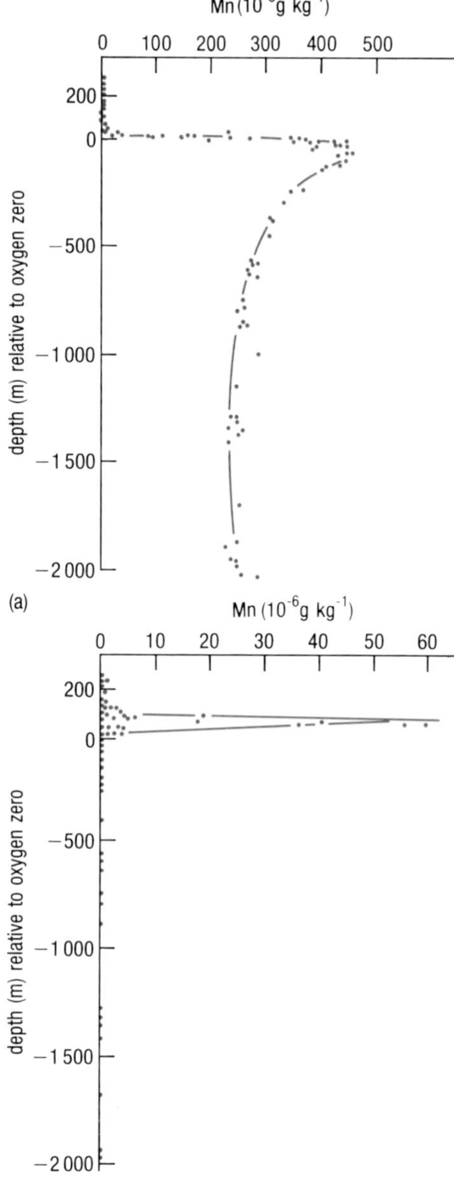

Figure 2.23 Concentration–depth profiles of manganese in the Black Sea, for use with Question 2.17. Note that the depth is relative to oxygen zero: water is oxygenated above the 0 mark, anoxic below. Note also that the concentration scales are such that the small variations in Figure 2.12(a) would not show up.

is greater than the available oxygen supply can cope with, the lower part of the water column may become completely **anoxic**.

In isolated basins such as the Black Sea and Norwegian fjords, density stratification of the water column and topographic barriers to mixing (sills at the mouths of basins) limit the supply of oxygen to the bottom waters by advection. The lower part of the water column thus becomes anoxic for at least a part of each year. Anoxic conditions can also be locally produced in coastal waters where human activities increase the supplies of nutrients and organic matter: examples include fish farms and pulp mills, as well as run-off of fertilizers from agricultural land.

Where the water is anoxic, other oxidizing agents must be used by bacteria to consume organic matter. Sulphate is a major dissolved constituent in seawater (Table 2.1) and when oxygen has been used up, one of the more favourable reactions (in terms of chemical energy), for the decomposition of organic matter is:

$$CH_2O + SO_4^{2-} = H_2S + HCO_3^- \quad (2.5)$$

As we saw earlier, the redox state of the water column can influence the solubility of trace elements that occur in more than one valency state. Such elements are sometimes said to be *redox-sensitive*. Manganese is one example of a redox-sensitive element. It is more soluble under reducing than under oxidizing conditions, as we saw in Section 2.2.3.

QUESTION 2.17 Explain which of the two profiles in Figure 2.23 represents dissolved manganese (Mn(II)) and which represents particulate manganese (Mn(IV)).

The decrease in manganese concentrations below oxygen zero in Figure 2.23(a) is probably due to the formation and precipitation of relatively insoluble $MnCO_3$, making use of bicarbonate ions provided by reaction 2.5 (precipitation of MnS is less likely because it is a relatively more soluble compound). In Figure 2.23(b), the concentration peak lies just above the interface between oxygenated and anoxic water. The size of the peak is only partly explained by precipitation, scavenging and sinking of manganese oxide particles in surface waters. An additional process is upward diffusion of water rich in dissolved manganese from below the interface, to be oxidized and precipitated in particulate form in the overlying oxygenated water. Similar relationships are found in sediments, where pore waters become anoxic at some depth below the sea-bed, as we shall see in Chapter 5.

Where the lower part of the water column is anoxic, however, as in Figure 2.23, the surface sediments will themselves be anoxic muds rich in organic carbon and associated trace elements. The example shown in Figure 2.24 is an open shelf area subject to seasonal upwelling and is a rather special case; but it serves to illustrate the general principle that elements which normally exhibit recycled profiles (Figure 2.13(b)) can accumulate to quite high levels in sediments where oxygen is deficient in the water column.

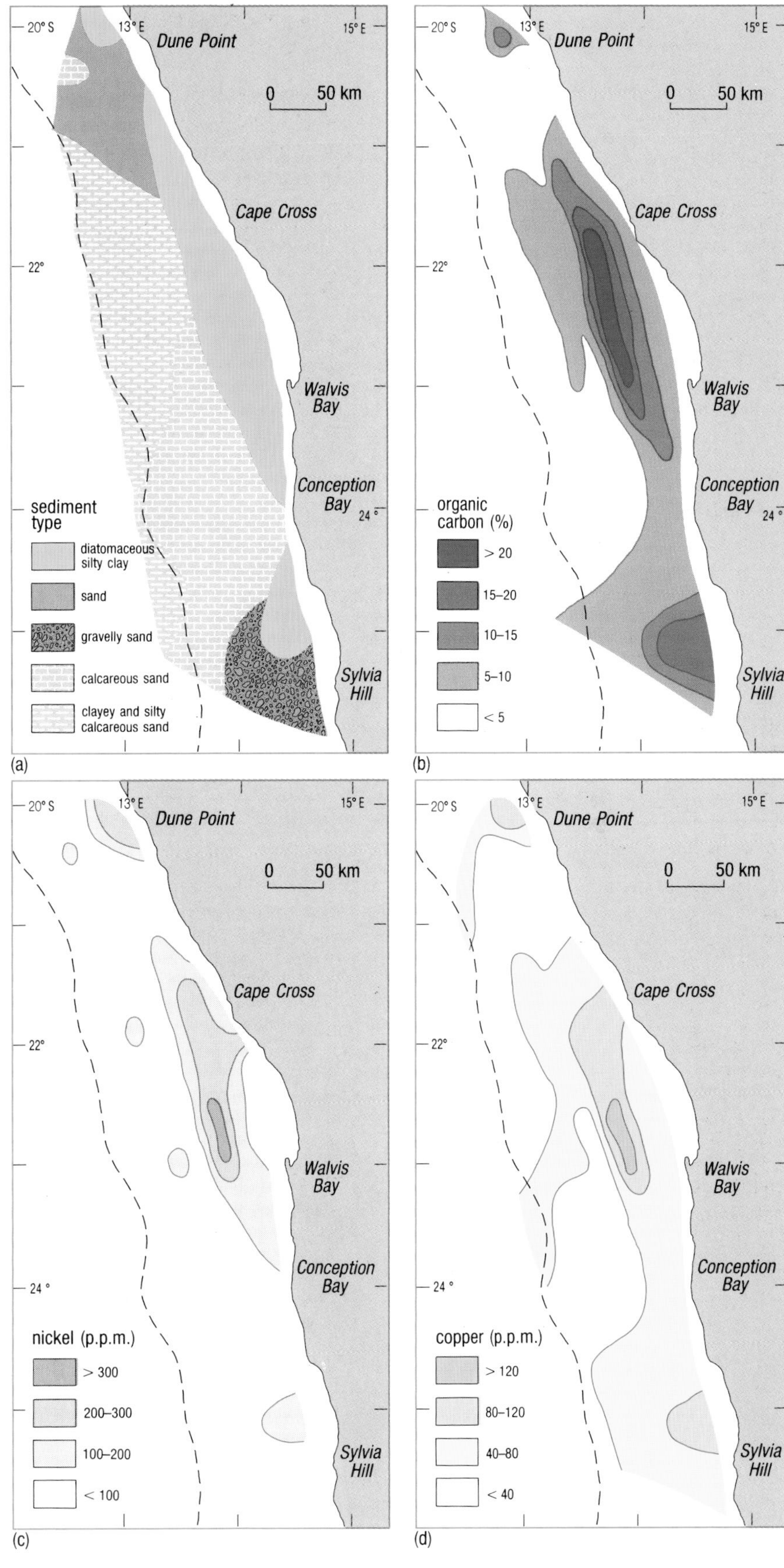

Figure 2.24 Maps of (a) sediment distributions, (b) organic carbon, (c) nickel and (d) copper concentrations in sediments of the continental shelf off south-western Africa, where upwelling occurs. Lead and zinc were also measured in this study, and show generally similar distribution to nickel and copper. The dashed line, the 500m depth contour, marks the approximate boundary of the continental shelf.

2.6 SUMMARY OF CHAPTER 2

1 Most naturally occurring elements have been detected in solution in seawater. Variations of salinity do not affect the overall constancy of composition of seawater with respect to major constituents, most of which behave conservatively. Most minor and trace constituents (and a few major ones, notably carbon, and calcium to a small extent) are non-conservative, because their concentrations are affected by biological processes in the oceans.

2 The oceans are in a steady state. There is an overall balance between the rate of supply of dissolved constituents (including excess volatiles and cyclic salts) and their rate of removal from solution. Residence times range from several tens of millions of years to a few hundred years or less, but most residence times are long compared with the oceanic mixing time. While in the oceans, dissolved constituents participate repeatedly in (mainly) biological cycles before being ultimately removed from solution in seawater into sediments and rocks at the sea-bed. Eventually, the sediments and rocks of the oceanic crust are accreted to continental margins or returned to the Earth's mantle by subduction at ocean trenches.

3 Much organic matter produced by primary production in surface waters is recycled there, but a proportion sinks out of the photic zone towards the sea-bed. Particulate organic matter ranges from micrometres to centimetres in size. Larger particles are mainly marine snow and faecal material, while the smallest sizes are dominated by bacteria. The composition of sinking particles changes with depth: skeletal material (carbonate and silica) dissolves only slowly, but organic matter is consumed and decomposed by animals and bacteria, and the residue becomes more refractory with depth as nutrients are extracted—Redfield ratios change in favour of carbon. Particulate organic carbon contributes only a small proportion of the organic carbon in seawater ($c.\,0.05$–$0.1\,\text{mgC}\,l^{-1}$ on average). Most organic carbon is in the form of dissolved organic molecules ($c.\,0.5$–$1\,\text{mgC}\,l^{-1}$ on average).

4 Particulate organic matter is a major regulator of seawater composition for minor and trace constituents in particular. Elements classified as recycled have concentration–depth profiles resembling those for the major nutrients (nitrate, phosphate, silica). They are taken up by organisms during growth and released back into solution (re-mineralized) as the organisms are consumed and decomposed on sinking into deeper waters after death. Recycled elements can also be subdivided into biolimiting (concentrations reduced to near-zero in surface waters), and bio-intermediate (concentrations only somewhat reduced in surface waters). The distinction between these two categories depends greatly on the amounts used in biological production relative to their total concentrations in seawater. Scavenged elements have profiles showing depletion at depth, the result of adsorption onto particle surfaces (mainly bacteria) and scavenging from the water column by larger sinking particles. The solubility of elements with more than one valency state can vary according to whether conditions are oxidizing or reducing.

5 The two-box model enables first-approximation estimates to be made of the relative amounts of dissolved constituents that are removed into sediments and recycled in the water column. It can only be applied to

elements in the recycled category. Biolimiting constituents are almost completely removed from solution in surface waters, but only a minute fraction of the particulate matter containing them reaches sediments on the sea-bed; the remainder is recycled, mainly above the permanent thermocline.

6 Lateral variations of biologically active constituents are characteristic of the deep oceans. The present-day pattern of surface currents and the deep circulation results in an overall enrichment of nutrients and other biologically active constituents in the deep waters of the North Pacific. The molar ratio of N:P in seawater is approximately the same as the Redfield ratio in organic tissue, at about 15:1.

7 Transfer of gases, liquids and solids takes place across the air–sea interface. Gas exchange occurs by molecular diffusion and is continuous, but at equilibrium there is no *net* flux in either direction. There is approximate equilibrium between atmosphere and ocean for the major gases. Minor gases produced by organisms in surface waters have a net flux from sea to air. Both gases and solids are also transferred in solution or suspension in water across the air–sea interface. In this case, the air–sea flux depends on the precipitation and on the washout ratio; and the sea–air flux depends on the extent to which aerosols are produced.

8 The deep oceans are well supplied with oxygen by deep water masses formed in polar regions. Dissolved oxygen concentrations decrease as water masses 'age' on moving away from their source regions and marine organisms use the oxygen in respiration. Maximum oxygen depletion occurs at about 1km depth, where the oxygen-minimum layer coincides approximately with maximum nutrient concentrations. Sub-oxic conditions develop in some oxygen-minimum layers. High levels of biological production can cause some coastal waters to become sub-oxic or anoxic; and in the absence of oxygen, bacteria use oxidizing agents such as sulphate to decompose organic matter. Sediments in anoxic muds can have high concentrations of organic carbon and trace metals.

Now try the following questions to consolidate your understanding of this Chapter.

QUESTION 2.18 The molar ratio of elements in particulate matter dominated by calcareous organisms is 130:25:15:1 (C:Ca:N:P). Why is there more carbon in this ratio than in the 'normal' Redfield ratio?

QUESTION 2.19 Samples of surface and deep water from the Atlantic and Pacific Oceans were analyzed for zinc and for cerium. Unfortunately, the samples were incompletely identified and it was necessary to inspect the data in order to determine which samples came from which ocean. Can you do the same, with the help of Figure 2.13 using the data which are tabulated below? In which of the two oceans were stations I and II located?

Table 2.5 For use with Question 2.19.

	Zinc (10^{-9} mol kg^{-1})		Cerium (10^{-12} mol kg^{-1})	
	Station I	Station II	Station I	Station II
Surface	0.8	0.8	66	19
Deep	1.6	8.2	11	6

QUESTION 2.20 How meaningful are the averages in Table 2.1, so far as the scavenged group of elements are concerned?

QUESTION 2.21 To what extent can the two-box model be applied to cyclic salts?

QUESTION 2.22 In making flux calculations for the transfer of DMS from sea to air, the value of ΔC in equation 2.1 is taken simply to be the measured concentration in surface seawater (see the end of Section 2.5.1). Why is that?

QUESTION 2.23 Which of the following statements are true, and which are false?

(a) All major constituents are conservative in seawater.

(b) The constancy of composition of seawater extends to recycled constituents.

(c) Marine snow makes up the largest proportion of particulate organic matter in the oceans.

(d) Considering the oceans as a whole, a greater volume of water is involved in upwelling than in downwelling.

(e) Continental shelves are kept well-oxygenated by deep water that has sunk and moved away from polar regions.

(f) The ratio of sulphate to total salinity, i.e. the ratio $SO_4^{2-} : S$, will be less in anoxic than in oxygenated seawater.

QUESTION 2.24 During 1988, high concentrations of toxic chemicals were found in sediments in an area of the North Sea where ships incinerated chemical wastes. The chemicals were products of incomplete combustion of chlorinated hydrocarbons, and were emitted from the stacks of the incinerator ships. Summarize *two* processes which ensured that these materials reached the sea-bed in the immediate vicinity rather than being spread over a wider area.

CHAPTER 3 THE ACCUMULATION OF PELAGIC BIOGENIC SEDIMENTS

Deep-sea sediments of biological origin occur in quite well-defined areas (Figure 1.4). The main factors controlling this distribution include the productivity of surface waters, the water depth, and the supply of terrigenous sediment, large volumes of which would dilute the biogenic components. The geographical separation of calcareous and siliceous sediments is related more to the different solubilities of calcium carbonate and silica and the chemistry of the water column than to the distribution of organisms at the surface.

Pelagic biogenic sediments consist mainly of the skeletal remains of very small pelagic organisms, some only a few μm in size (Section 1.1.1). Stokes' law calculations suggest that these remains would take decades or even centuries to sink to the deep sea-floor. Such calculations assume that the particles have spherical shapes without protrusions, and that there is no turbulence in the water column, neither of which applies in the oceans; so, the time taken for such material to reach the sea-bed in the deep oceans would be even longer.

So, how is it that sediments formed of this skeletal debris mainly occur directly below areas of high productivity in surface waters? Why are they not spread all over the sea-bed by ocean currents?

As you read in Chapter 2, the short answer is biopackaging, either in the form of marine snow or faecal pellets—it has been estimated that a single faecal pellet can contain as many as 10^5 coccoliths, so biological aggregation is obviously an important way of speedily transferring planktonic debris (along with pelagic clays of mainly aeolian origin) to the sea-bed.

3.1 THE PRESERVATION OF PELAGIC CARBONATES

Rapid descent through the water column is only the first step towards the conversion of calcareous skeletal material into carbonate sediment at the sea-bed. The chemistry of deep ocean waters determines whether or not this conversion occurs.

In general, productivity exerts a greater influence on the actual composition of pelagic carbonates than does depth, because Foraminifera increase faster than coccolithophores in the photic zone as overall productivity rises. As a result, calcareous nanofossil oozes tend to predominate in areas of low productivity, while foraminiferal oozes are found mainly where productivity is high.

3.1.1 CARBONATE SATURATION IN SEAWATER

Carbon dioxide gas is more soluble in cold water than in hot water, and its solubility increases with pressure. This property is well known to those manufacturers and consumers of bottled and canned drinks which froth or fizz when opened. The CO_2 combines with water molecules to produce a weak acid (carbonic acid) which then dissociates to produce hydrogen and bicarbonate ions:

$$CO_2 \text{ gas} + H_2O = H_2CO_3 = H^+ + HCO_3^- \tag{3.1}$$

As noted in Section 2.5.1, the concentration of carbon dioxide as *gas* in seawater is very small. In surface waters, only about 1 atom of carbon in 200 is in the form of dissolved CO_2 molecules, and even in the deep ocean the figure rises only to about 3 atoms in 200.

There is a further component of reaction 3.1:

$$HCO_3^- = H^+ + CO_3^{2-} \tag{3.2}$$

Carbon thus occurs as several species in solution: CO_2 gas, H_2CO_3, HCO_3^-, and CO_3^{2-}, as well as carbon combined in organic molecules (which rarely amounts to more than about 1 p.p.m., Section 2.2.1). HCO_3^- and CO_3^{2-} are quantitatively by far the most important of these. Reaction 3.2 is rapid, and seawater can be assumed to contain an equilibrium mixture of the three ions.

A large proportion of the bicarbonate and carbonate ions in seawater comes not from direct solution of atmospheric carbon dioxide, however, but from rivers flowing into the sea: the weathering of rocks by carbonic acid in rainwater releases cations (e.g. Ca^{2+}, Na^+, K^+) and bicarbonate and carbonate ions (along with other constituents, of course) into solution in river water. The average concentration of bicarbonate plus carbonate ions in river water is around $60 \text{mg} l^{-1}$, which is equivalent to about $12 \text{mg} l^{-1}$ (12 p.p.m.) of carbon, nearly half the figure for the average concentration of carbon in seawater (Table 2.1).

Figure 3.1 shows typical profiles for total dissolved inorganic carbon in seawater, expressed as ΣCO_2 (where Σ is capital sigma, and denotes 'sum of'). The profiles illustrate how carbon in its various forms is the least conservative of the major dissolved constituents, with obvious bio-intermediate character (*cf.* Figure 2.13). Concentrations are higher in the Pacific than in the Atlantic, which is consistent with the pattern described for recycled constituents in Section 2.4.

Profiles of total dissolved inorganic carbon (ΣCO_2) in seawater take the form shown in Figure 3.1 chiefly because carbon dioxide is removed from solution in surface waters by photosynthesis (reaction 2.3 goes to the right) and returned to solution in deep water, as organic matter is decomposed (reaction 2.3 goes to the left). Another contribution to the increase of ΣCO_2 with depth comes from the dissolution of atmospheric carbon dioxide at high latitudes in cold surface waters, which sink to the deep sea-floor on account of their low temperature and increased density.

In what follows, it is essential to bear in mind the point made in Section 2.5.1 and above, that nearly all carbon dioxide in seawater is in the form of bicarbonate and carbonate ions, because of reactions 3.1 and 3.2.

For our purposes, it is sufficient to state that the equilibrium relationships in reactions 3.1 and 3.2 are such that, as ΣCO_2 increases, so does the ratio of bicarbonate to carbonate ions in the expression:

$$[H^+] = K \frac{[HCO_3^-]}{[CO_3^{2-}]} \tag{3.3}$$

where the terms in square brackets are molar concentrations and K is the equilibrium constant for reaction 3.2.

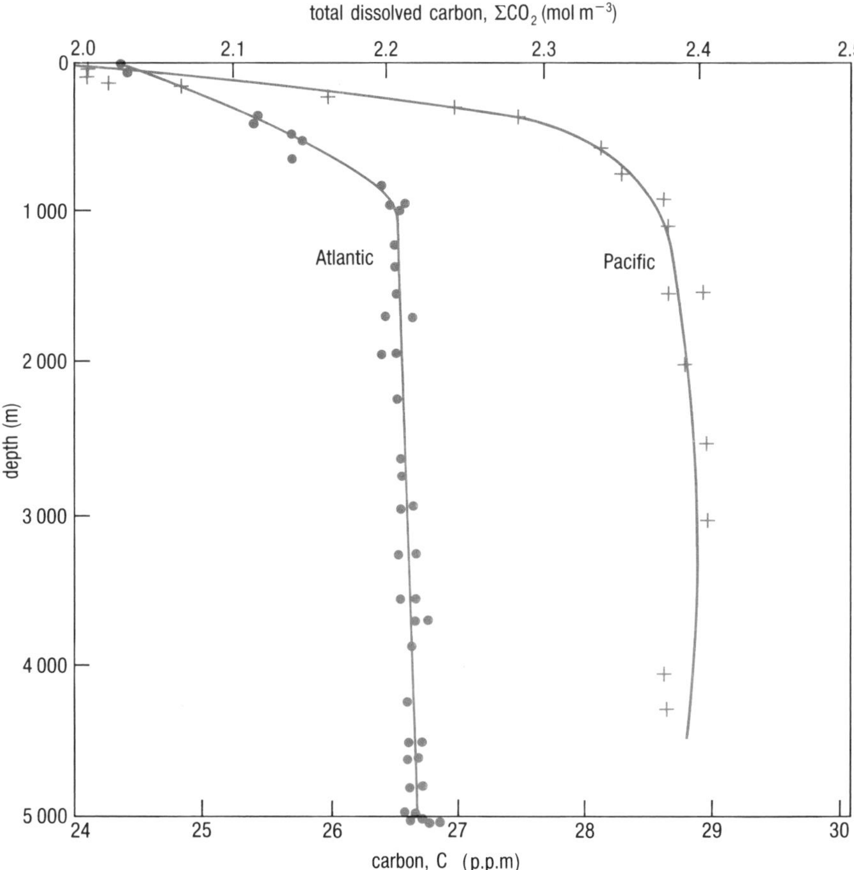

Figure 3.1 Variation with depth of total dissolved carbon (ΣCO_2), in the Atlantic (36°N, 68°W) and in the Pacific (28°N, 122°W). Note that Σ (sigma) denotes 'sum of'.

If ΣCO_2 increases, will the water become more or less acid according to equation 3.3?

As ΣCO_2 increases, the ratio of bicarbonate to carbonate ions increases and so does $[H^+]$; i.e. there are more hydrogen ions, and the water becomes more acid (pH decreases). The converse will be true where ΣCO_2 decreases. Thus, variations in ΣCO_2 significantly affect the balance between bicarbonate and carbonate ions.

However, because of the rapidity of reaction 3.2, this component of the carbonate system provides a chemical buffer for pH in the oceans. Considerable changes of $[\Sigma CO_2]$ or of the $[HCO_3^-]$:$[CO_3^{2-}]$ ratio (or even of $[H^+]$) are required before the pH is significantly changed.

Dissolution of calcium carbonate skeletons ($CaCO_3$) occurs where increased acidity of the water results from the release of hydrogen ions:

$$CaCO_3 + H^+ \rightarrow Ca^{2+} + HCO_3^- \qquad (3.4)$$

Therefore, is $CaCO_3$ more likely to dissolve where ΣCO_2 concentrations are high than where they are low?

It must follow from what you have read that where ΣCO_2 concentrations are high, then so is $[H^+]$, the water is more acid, so $CaCO_3$ is more likely to dissolve.

Accordingly, would you expect $CaCO_3$ to dissolve more readily in Atlantic than in Pacific waters, judging from Figure 3.1?

ΣCO_2 is higher in the Pacific than in the Atlantic, so $CaCO_3$ should dissolve more readily in Pacific waters.

However, Figure 1.4 indicates also that the solubility of calcium carbonate in the oceans is depth-dependent: carbonate sediments are abundant on shallower parts of the ocean floor, notably the mid-ocean ridges, but absent from the deeper abyssal plains. Calcium carbonate is more soluble in cold than in warm water and it is also more soluble at high pressure than at low pressure (dissolved calcium and carbonate ions occupy less volume than when combined in solid form).

QUESTION 3.1 Does that help you to explain part of the answer to Question 1.1, namely: why do calcareous biogenic sediments predominate along the ocean ridges?

But that is not the whole story: another expression for the dissolution of calcium carbonate is:

$$CaCO_3 \rightleftharpoons Ca^{2+} + CO_3^{2-} \qquad (3.5)$$

In the oceans, the concentration of Ca^{2+} in solution is virtually constant. As you read in Section 2.1, calcium in seawater departs only very slightly from conservative behaviour. The concentration of carbonate ions, i.e. $[CO_3^{2-}]$, is much more variable.

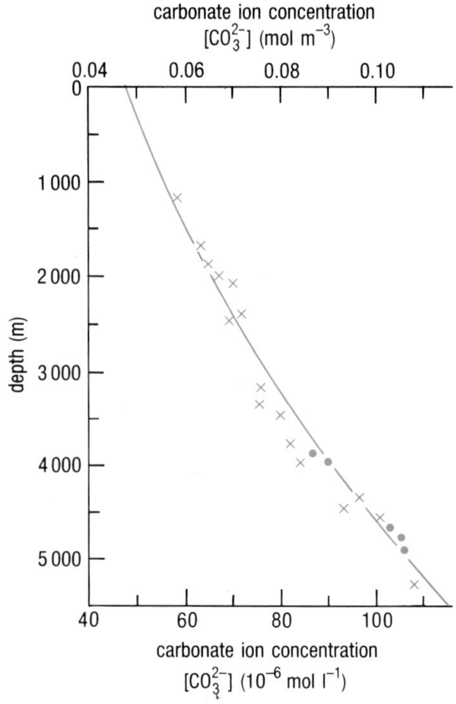

Figure 3.2 The saturation curve of CO_3^{2-} concentration versus depth for calcite, $CaCO_3$. The theoretical relationship is given by the solid line; crosses are experimentally determined points; dots represent actual observations of calcite dissolution in the oceans.

Figure 3.2 shows how the concentration of CO_3^{2-} ions in seawater determines whether the calcite variety of calcium carbonate (Section 1.1.1) will be dissolved or not. If the concentration of CO_3^{2-} ions in the seawater lies to the right of the line, calcite will not dissolve. If $[CO_3^{2-}]$ lies to the left of the line, calcite will dissolve.

In order to predict the depth at which calcite skeletal material will begin to dissolve in the water column, therefore, all we need to do is to determine $[CO_3^{2-}]$ in the water, and plot the value in Figure 3.2.

Unfortunately, $[CO_3^{2-}]$ cannot be directly measured. It has to be determined by indirect means. The concentration of ΣCO_2 in a seawater sample can be found easily by direct analysis. The sample can also be titrated with acid, to convert bicarbonate and carbonate to CO_2:

$$HCO_3^- + H^+ \rightarrow CO_2 + H_2O \qquad (3.6)$$

$$CO_3^{2-} + 2H^+ \rightarrow CO_2 + H_2O \qquad (3.7)$$

The result of this titration gives a value for the **alkalinity** of the sample, which for our purposes can be defined as the combined concentrations of bicarbonate and carbonate ions, expressed in 'charge-equivalent' terms. If we express concentrations in molar terms, then we can write:

$$[\Sigma CO_2] = [HCO_3^-] + [CO_3^{2-}] \qquad (3.8)$$

$$\text{Alkalinity, } A = [HCO_3^-] + 2[CO_3^{2-}] \qquad (3.9)$$

By re-arranging these two equations:

$$A - [\Sigma CO_2] = [CO_3^{2-}] \qquad (3.10)$$

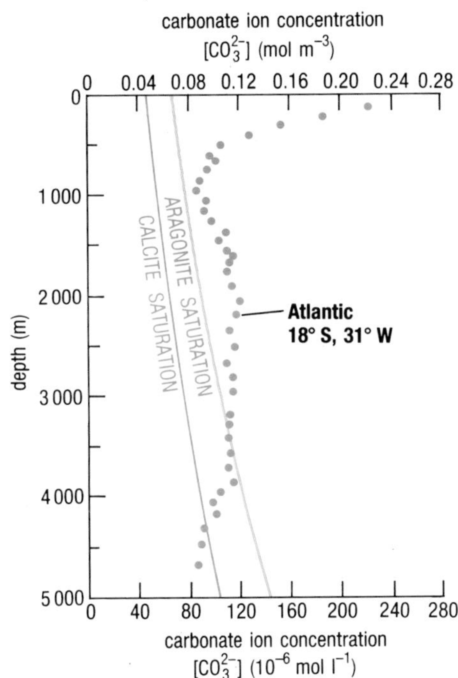

Figure 3.3 Saturation curves of CO_3^{2-} concentration versus depth for calcite and aragonite (both $CaCO_3$). The curve for calcite is from Figure 3.2. Dots represent a profile of calculated values of [CO_3^{2-}] for a station in the Atlantic.

Thus, by direct measurement of ΣCO_2 and A, the value of [CO_3^{2-}] can be calculated. In practice, for accurate results it is a little more complicated than this simple outline suggests, because as we have seen, bicarbonate and carbonate are not the only contributors to total dissolved inorganic carbon (ΣCO_2) in seawater; and other dissolved species contribute to alkalinity. But our approach violates no basic principles and gives adequate approximations of the real situation.

QUESTION 3.2 The alkalinity of a sample of surface water is 2.35 mol m^{-3} and its [ΣCO_2] is 2.15 mol m^{-3}. The same quantities for a sample of water from 4 km depth at the same location are 2.45 mol m^{-3} and 2.40 mol m^{-3} respectively. Work out the [CO_3^{2-}] for each sample, using equation 3.10. Where would these samples plot in relation to the line in Figure 3.2, and what does that tell you about the degree of saturation of the water samples with respect to calcite?

Some calcium carbonate skeletal material is formed of aragonite rather than calcite, and aragonite is less stable than calcite and dissolves more readily (Section 1.1.1). Figure 3.3 shows the same curve for calcite as in Figure 3.2, along with the analogous curve for aragonite, and an actual profile for [CO_3^{2-}] from an Atlantic station.

At what depth(s) would you expect calcite and aragonite skeletal material to begin dissolving at this station?

Figure 3.3 suggests that aragonite should begin to dissolve at a little over 3 km depth, whereas calcite should not start dissolving till nearly 4.5 km depth.

An important general point is also illustrated in Figure 3.3: down to a few km depth, seawater is supersaturated with respect to calcium carbonate (*cf.* Question 3.2). The degree of supersaturation is greater for calcite than for aragonite (the saturation curve for aragonite lies to the right of that for calcite). The reason why $CaCO_3$ does not precipitate spontaneously from seawater is that most of the CO_3^{2-} ions in the solution are 'combined' with Mg^{2+} ions in what are known as *ion pairs*, which must be extensively broken up for $CaCO_3$ to precipitate spontaneously.

3.1.2 THE LYSOCLINE AND THE CARBONATE COMPENSATION DEPTH

The rate of descent of rapidly sinking particles generally takes most of the skeletal material through the undersaturated lower parts of the water column in too short a time for significant dissolution to occur. *Dissolution of calcareous material takes place mostly at the sea-bed.*

The depth at which the dissolution of carbonate skeletal material is observed to begin, from observation and analysis of sediment samples from the ocean floor, is called the **lysocline**. Below the lysocline, dissolution occurs at increasing rates, so that there is a progressive decrease in the proportion of carbonate skeletal material preserved in the sediments. The depth at which this proportion falls below 20% of the total sediment is called the **carbonate compensation depth, CCD**.

In theory, the depth of the lysocline, based on visual inspection of carbonate debris in sediments collected at different depths, should coincide with the depth predicted from intersections of [CO_3^{2-}]-depth profiles with the saturation curves, as in the example given in Figure 3.3. In practice, the (observed) lysocline may coincide with the predicted depth, but can be as much as a few hundred metres below it.

Figures 3.2 and 3.3 show that seawater becomes progressively more undersaturated with respect to calcium carbonate (whether calcite or aragonite) as depth increases. In consequence, the rate of dissolution of carbonate increases with depth below the lysocline.

The depth of the CCD is controlled in part by how undersaturated the water is, but also by the flux of calcareous debris to the sediments. The greater this is, the greater the likelihood of carbonate material being buried before it dissolves. Figure 3.4 shows the depth of the CCD for calcite, as determined from deep-sea sediment samples. The CCD is depressed beneath the Equator on account of the high productivity resulting from equatorial upwelling, and the increased flux of calcareous remains to the sea-bed.

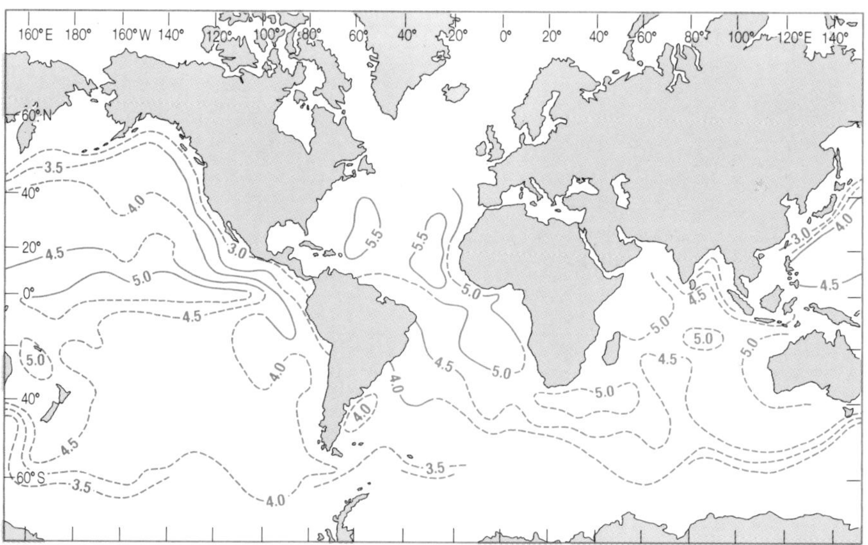

Figure 3.4 Contours (in km) for the calcium carbonate compensation depth (CCD), defined by interpolating boundaries between calcareous sediments and sediments with little or no calcium carbonate. Solid blue contours represent more than 20 control samples per 10° square; broken blue contours represent fewer than 20 control samples. Note that this map is for the more common calcium carbonate mineral, *calcite*; the compensation depth for the other variety, *aragonite*, is very different, as discussed in the text.

At this point, it is necessary to emphasize two things:

1 The lysocline is a 'surface' that can be 'mapped' within the oceans by reference to the chemistry of the water column (Figures 3.2 and 3.3). That level can be *checked* by inspection of calcareous sediments accumulating on the sea-bed at appropriate depths: if no dissolution is observed, then the sediments are above the lysocline; if dissolution has begun, the sediments are below the lysocline.

2 The CCD is also 'a surface', but it can be 'mapped' *only* by inspection and analysis of calcareous sediment samples from the sea-bed. If originally calcareous sediments contain more than 20% $CaCO_3$, they are above the CCD; if they contain less, they are below it. On maps of the CCD such as Figure 3.4, the contour lines are *interpolations* between sea-bed locations where the CCD has been determined from inspection of the sediments.

Figure 3.4 shows that the level of the CCD typically rises towards the continental margins, where biological productivity is in general greater

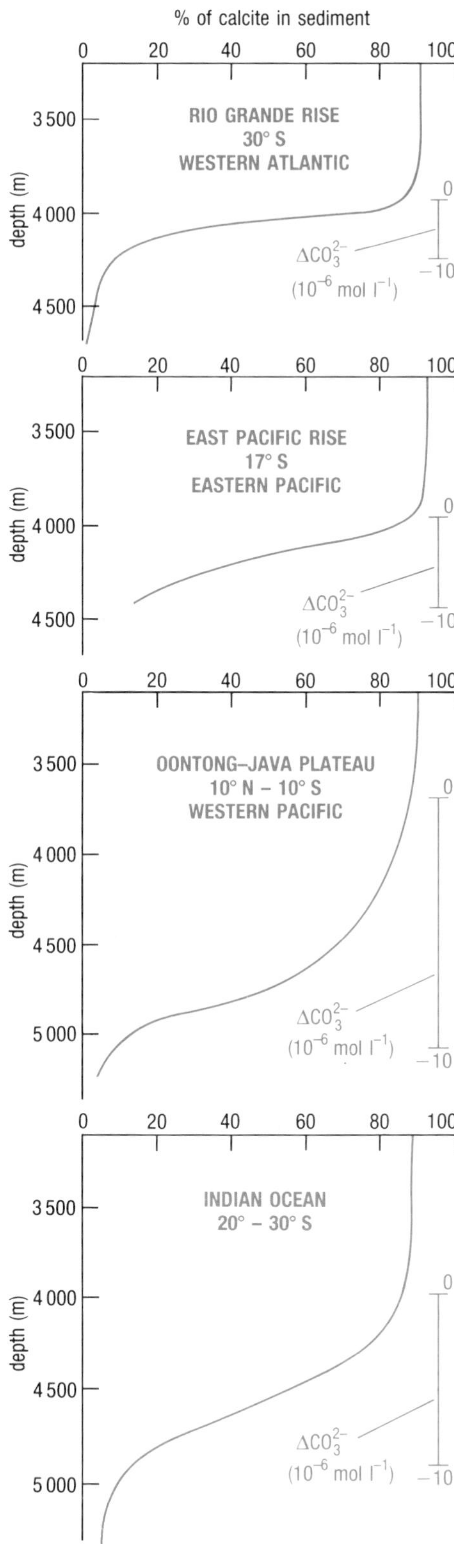

Figure 3.5 Generalized profiles of the calcite content of surface sediments versus water depth on the flanks of ocean ridges and plateaus in different parts of the ocean. For explanation of ΔCO_3^{2-}, see the text. For use also with Question 3.3. Note that 10×10^{-6} mol l^{-1} = 0.01 mol m^{-3}.

than throughout most of the open oceans. This would seem to contradict the correlation between high productivity and depression of the CCD which we have just established. The reason for the apparent contradiction is that dissolution of calcareous debris is aided by a rich supply of organic matter which is available for consumption by benthic animals and bacteria at the sea-floor, releasing CO_2 into solution at the sediment–water interface. The CO_2 instantly combines with the water, as in reaction 3.1, increasing the ΣCO_2 of the bottom water, which thus becomes more acid and more undersaturated in CO_3^{2-} with respect to calcite (*cf.* equation 3.3).

In these situations, the CCD virtually coincides with the lysocline, whereas in deeper water there is a depth interval of several hundred metres between them. This is summarized in Figure 3.5, which shows how the amount of calcite in sediments decreases with depth in different parts of the oceans.

QUESTION 3.3 You know that the lysocline is the depth at which dissolution commences (i.e. where the percentage of $CaCO_3$ in the sediments begins to decrease); and that the CCD is the depth at which the proportion of $CaCO_3$ in the sediments falls to less than 20%. Hence, identify the lysocline and the CCD on the profiles on Figure 3.5. Which of the two levels shows the greater variation in depth?

Also shown in Figure 3.5 is the depth interval over which the *difference* between the saturation concentration of CO_3^{2-} for calcite and the actual concentration of CO_3^{2-} in the water rises from 0 to 10 (in 10^{-6}moll^{-1}). This difference is expressed as ΔCO_3^{2-} and on Figure 3.5 it is shown to range from 0 to −10. The rate of dissolution of calcite increases below the lysocline, as the value of $-\Delta CO_3^{2-}$ increases, i.e. as the water becomes more and more undersaturated with respect to calcite. You can see from Figure 3.5 that undersaturation of CO_3^{2-} of only 10×10^{-6}moll^{-1} is enough to dissolve virtually all the calcite in the sediments.

The reason for the transition zone between the lysocline and the CCD (the sloping part of each profile in Figure 3.5) is that the pore waters in the uppermost few millimetres of the sediment become saturated with respect to calcite. Dissolution can only continue when they become undersaturated again, through exchange with bottom waters (by diffusion and/or advection). The greater the value of $-\Delta CO_3^{2-}$, the longer it takes for saturation to occur and the easier it is for exchange with bottom waters to restore the undersaturation.

In general, the greater the flux or 'rain rate' of calcium carbonate debris, the greater the thickness of this transition zone: the time taken to restore undersaturation in the pore waters is increased where the supply of calcite debris is greater. So, the CCD is depressed beneath areas of high productivity, as we have seen.

You might well be tempted to ask at this point why the term 'lysocline' is not applied to the sloping part of profiles such as those in Figure 3.5, by analogy with the terms 'thermocline' and 'halocline'. This is a good question. It is likely that in the original definition, the lysocline *was* the depth interval over which dissolution of calcium carbonate occurred. Nowadays, however, the definition is generally confined to the depth at which dissolution *commences*.

The carbonate system in the oceans is one of considerable complexity, and we have been able to consider only some aspects. The basic relationships are really quite simple however:

1 The solubility of calcium carbonate in seawater increases with depth (e.g. Figure 3.2)

2 In general, the concentration of ΣCO_2 increases with depth (e.g. Figure 3.1), so that at some depth the seawater becomes sufficiently acidic and undersaturated with respect to calcium carbonate for it to begin to dissolve (e.g. Figure 3.3). This is the lysocline.

3 At some greater depth, nearly all the calcium carbonate has dissolved. This is the carbonate compensation depth, the CCD (e.g. Figure 3.5).

4 Calcite is more stable than aragonite, so the lysocline and CCD are both shallower for aragonite than for calcite (e.g. Figure 3.3).

Some of the complexities in the system arise from the inherent variability of the contributions to total dissolved inorganic carbon (ΣCO_2). At the start of Section 3.1.1, you read that some of the ΣCO_2 comes from the atmosphere: CO_2 gas goes into solution and combines with water (reactions 3.1 and 3.2). Gases dissolve more readily in cold than in warm water, so the deep water masses sinking in polar latitudes carry dissolved CO_2 to the ocean depths.

We have seen that as water masses 'age' on moving away from the surface source regions, their initial complement of oxygen is depleted by respiration and bacterial decomposition. The oxygen is used to form CO_2, which increases the concentration of ΣCO_2. There are thus considerable lateral and vertical variations of ΣCO_2. As exemplified in Figure 3.1, older Pacific (and Indian) Ocean waters carry more ΣCO_2 than younger Atlantic waters. Nor is the rate of decrease of $[CO_3^{2-}]$ regular with depth. In Figure 3.3 for example, there is a distinct minimum in the $[CO_3^{2-}]$ profile at about 1 km depth, superimposed on the overall downward decrease.

QUESTION 3.4 Can you offer an explanation for this minimum, at about 1 km depth?

The average level of the CCD is an indicator of the rate of removal of atmospheric CO_2 to the deep sea, both in organic tissue, and in carbonate skeletal material. As we noted earlier, there are two principal ways in which the CO_2 gets into the ocean in the first place. One is direct, through solution of CO_2 from the atmosphere. The other is indirect, through weathering of rocks on land (by carbonic acid in rainwater); this supplies bicarbonate and carbonate ions in river water and thence to seawater. These two routes for the entry of carbon into the oceans have different effects on the carbonate system there.

Reactions 3.1 and 3.2 show that direction solution of atmospheric CO_2 produces hydrogen ions to balance the negative charges on the bicarbonate and carbonate ions. In contrast, the bicarbonate and carbonate ions in solution in river water are balanced chiefly by cations such as Ca^{2+}, Na^+, K^+ and so on, rather than by H^+.

If the rate of supply of CO_2 to the oceans were to increase by either of these routes, which would be more likely to make seawater more acid?

Direct solution of CO_2 is accompanied by the formation of hydrogen ions in seawater, which would make the seawater more acid. The more CO_2 entering the oceans by this route, the greater the value of $[H^+]$, and the higher the ratio of bicarbonate to carbonate ions (equation 3.3). On the other hand, the bicarbonate and carbonate ions entering the oceans from rivers are mostly accompanied by cations other than H^+, so seawater does not get more acid (there are exceptions, of course, because some river waters are relatively acid, but the generalization is valid enough for our purposes). The concentration of hydrogen ions $[H^+]$ in seawater is not increased and the ratio of bicarbonate to carbonate ions is not affected. Indeed, $[H^+]$ might even fall, in which case the bicarbonate:carbonate ratio would also fall. Because of this, marine chemists sometimes speak of weathering (of rocks on land) as supplying alkalinity to the oceans—from equation 3.10, if $[CO_3^{2-}]$ is high, then A must also be high.

Do not worry if you have not been able to follow all the details of these arguments; it is more important to grasp the overall implications which are:

1 If ΣCO_2 in seawater is increased by direct solution of CO_2 from the atmosphere, the oceans will become more acid, $[CO_3^{2-}]$ in deep water will decrease, and the levels of the lysocline and the CCD will rise.

2 If ΣCO_2 in seawater is increased by the addition of bicarbonate and carbonate ions from rivers, the oceans will probably not become more acid (and might even become less so). $[CO_3^{2-}]$ in deep water will not change significantly (it might even increase), and the levels of the lysocline and CCD will not change (they might even fall).

Both of the scenarios summarized in 1 and 2 above are sensitive to changes in climate and biological activity, which are themselves linked to some extent and which have changed through geological time, with marked effects on the level of the CCD, as we shall see in the next Section.

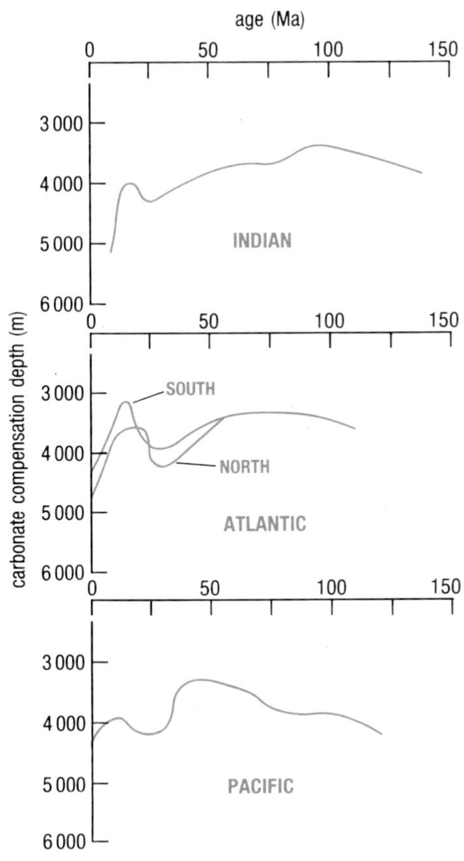

Figure 3.6 Changes in the level of the CCD with time for the major oceans, determined from sediment sequences recovered in deep-sea drill cores.

3.1.3 VARIATIONS OF THE CCD WITH TIME

The global average CCD appears to be deeper now than at any time in the past 150Ma (Figure 3.6). The CCD seems to have deepened progressively, if somewhat erratically, to its present level, over the past 100Ma.

Past levels of the CCD are worked out using information from deep-sea sediment cores and from the basic **age–depth relationship** of oceanic crust. The depth to the top of the igneous oceanic crust (below the sediments) increases with time, though at an exponentially decreasing rate (Figure 3.7), because of cooling, contraction and subsidence of oceanic lithosphere as it moves away from the ridge crest at which it formed. The 'ideal' or theoretical age–depth curve (the dashed line in Figure 3.7) is used to work out past levels of the CCD using information obtained from deep-sea sediment cores. The procedure is very simple in principle.

In any deep-sea sediment core taken from a site that formerly underlay a region where biogenic sediments were not 'swamped' by terrigenous inputs, there will be an upward transition from carbonate-rich sediments (below) to carbonate-poor or carbonate-free sediments (above). The age of the transition can be found easily from the fossil remains in the sediments. If this transition occurred, say, 40Ma ago, then according to

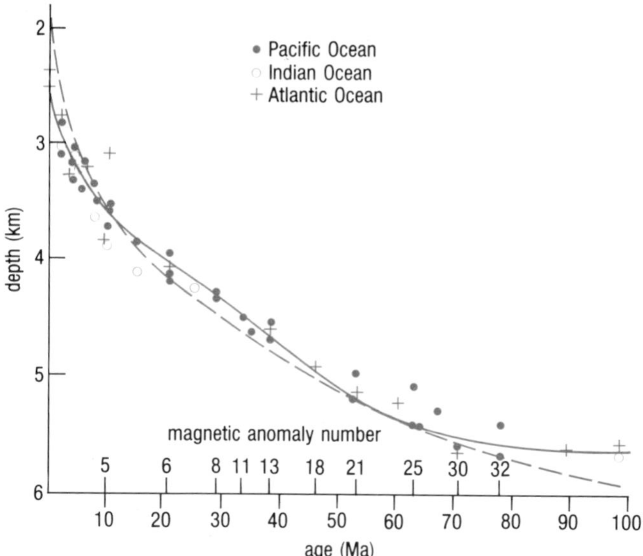

Figure 3.7 Observed and theoretical relationship between the depth to the top of the igneous oceanic crust and its age. The solid line is a best-fit curve through observed points. The dashed line is a theoretical elevation curve, calculated on the assumption that the increase of depth with age is due to thermal contraction of the lithosphere as the plate cools on moving away from the ridge axis. Magnetic anomaly numbers refer to the linear magnetic stripes on the ocean floor, which are arranged symmetrically about ridge axes.

the age–depth relationship (Figure 3.7), the top of the underlying igneous crust lay at a depth of about 4800m. Knowing the thickness of sediment below the transition, and making corrections for sediment compaction and isostatic depression due to sediment loading, the level of the CCD 40Ma ago can be determined, sometimes to within a couple of hundred metres.

It is one thing to establish a pattern, another to explain it. It would seem that in the Cretaceous the oceans were relatively more acid than at present, for the CCD to have been so much shallower (Figure 3.6). The Earth as a whole was warmer then, with ice-free poles. About 100Ma ago, as the continents dispersed and the ocean basins developed their present configurations, the world saw one of the greatest marine **transgressions** in its history, a vast inundation of low-lying land areas, the result of a global sea-level rise of some 300–400m. The planktonic organisms that secrete calcareous tests (coccolithophores and Foraminifera) did not begin to become widespread until about 100Ma ago. Before that time, carbonate sedimentation was mainly confined to the shallow waters of continental shelf regions; and so relatively less calcium was precipitated from open ocean waters than is the case now.

A relative scarcity of pelagic carbonates might have been a contributory factor in maintaining a shallow CCD, but it cannot have been the only one. Cretaceous sediment sequences containing dark, clay-rich layers with between 1% and 30% of organic carbon, offer another explanation. The oceans must have been poorly oxygenated for much of the period covered by Figure 3.6, because ice-caps and glaciers did not begin to develop until late in the Tertiary era. In the absence of ice-caps, there could be no regular supply of abundant cold, dense, well-oxygenated water to the deep oceans, as there is now. At the same time, the warm climatic conditions encouraged biological production, not only in the oceans themselves, but also on the now very extensive continental shelves and on land (the Cretaceous was a time not only of great inundation, but also of great increases in the number and diversity of land plants). The huge amounts of organic carbon supplied to the oceans were greater than the meagre supply of oxygenated water could cope with. There was widespread depletion of dissolved oxygen throughout the world's oceans, extending from a few hundred metres below the surface to about 2000–3000m depth. Carbon-rich sediments were deposited where these anoxic waters intersected the sea-bed. The waters were anoxic because the oxygen had all been used up in respiration and converted to CO_2. This would in turn have made the water sufficiently acid to dissolve calcium carbonate sinking from the surface, thus maintaining the shallow CCD.

As sea-level began to fall again after the Cretaceous, the amount of organic matter supplied to the oceans decreased as the areas of continental shelf became smaller. The deep oceans began to be supplied with oxygenated water from polar regions, and the CCD began to deepen towards its present level, albeit with some fluctuations.

By way of a postscript to this tale, it is worth mentioning why the record in Figure 3.6 extends back only about 150Ma. There is hardly any ocean floor older than this left in the ocean basins (Section 1.1). The floors of older oceans have nearly all been subducted back down into the Earth's mantle (Figure 2.2).

3.2 THE PRESERVATION OF PELAGIC SILICEOUS REMAINS

Broadly speaking, similar factors affect the accumulation of siliceous sediments as affect the accumulation of carbonate sediments; the supply of skeletal debris, the extent to which dissolution is able to take place and whether or not the siliceous debris becomes diluted on the ocean floor by other types of sediment. There is also a fundamental difference affecting the preservation of the two types of sediments; seawater is everywhere undersaturated with respect to SiO_2, whereas only deep waters are undersaturated with respect to $CaCO_3$.

The solubility of amorphous (opaline) silica decreases by about 30% for a fall in temperature from 25 to 5°C, though this decrease is offset somewhat in the deep oceans, because high pressure acts to increase the solubility slightly. Inspection of samples from sediment traps suggests that most dissolution of opaline silica occurs at the sea-bed, although profiles such as Figure 2.9(c) testify to considerable dissolution and recycling in upper parts of the water column.

Siliceous skeletal material that reaches the sea-bed must be rapidly transported in large, fast-sinking particles (marine snow and faecal pellets). The greater the supply of skeletal debris, the more likely some of it is to reach the ocean floor. Supply is related to productivity, and siliceous sediments on the ocean floor thus underlie areas of high productivity (Figure 1.4). Even in these areas, however, only between 1 and 10% of siliceous material escapes dissolution, either in the upper parts of the water column or at the sea-bed, and accumulates to form sediments. This small proportion can be easily swamped, especially where there is a high input of terrigenous sediment, or where the sea-floor is above the CCD or lysocline.

As in the case of carbonate, the pore waters in the uppermost few millimetres of the sediment become saturated with respect to silica, and undersaturation is only restored by exchange with bottom waters. Clearly, the greater the flux of siliceous debris, the more difficult it is for undersaturation to be restored and the better the chances of preservation of the siliceous remains.

Close to continental margins where inputs of terrigenous material are high, sediments can accumulate at rates as high as a few metres per *thousand* years, enough to obliterate any biogenic component. By contrast, the accumulation rate of a siliceous ooze is of the order of a few metres per *million* years; whereas accumulation rates of pelagic carbonates can reach several tens of metres per million years, which is also quite enough to dilute the siliceous skeletal content of sediments beneath regions of high productivity, where the CCD is depressed.

As recorded in Chapter 1, diatom oozes generally predominate below the CCD in high latitudes and in areas of coastal upwelling, whereas radiolarian oozes occur in tropical regions. The belt of siliceous sediments round the Antarctic (Figure 1.4) consists of diatom ooze, whereas that in the equatorial Pacific is of radiolarian ooze. Mixed radiolarian and foraminiferal oozes may occur in the most productive areas, where there has been some depression of the CCD.

3.3 BIOGENIC SEDIMENTS AND PALAEOCEANOGRAPHY

Sediments recovered in deep-sea drill cores contain a wealth of information about past climates and changing patterns of ocean circulation.

Calcareous sediments can provide information about more than past levels of the CCD (Section 3.1.3). Different species of organisms inhabit different water masses of contrasted temperature, salinity and other properties. For example, modern benthic foraminiferal assemblages show a correlation with present-day oceanic bottom water masses. Studies of similar assemblages of benthic Foraminifera in deep-sea sediment cores should therefore enable the past distribution and extent of bottom water masses to be mapped, *always assuming that the species concerned have not changed their environmental requirements with time.*

Figure 3.8 illustrates some results from one such investigation, which suggests that the influence of Antarctic Bottom Water (AABW) may have extended as far as the British Isles during mid- to late Miocene times, but no further than the latitude of Gibraltar since the late Pliocene. The southwards 'retreat' of the AABW appears to have been accompanied by intensified production of North Atlantic deep and bottom water masses in the Pliocene and Pleistocene. Correlation with data from other regions suggests that maximum build up of Antarctic ice occurred in the late Miocene, around 5Ma ago, while the development of major Arctic ice-sheets took place in the Pliocene, some 3–2.5Ma ago.

There is another way in which calcareous organisms can provide information about past conditions in the oceans. This is through the ratio of the two main isotopes of oxygen, ^{16}O and ^{18}O, in the calcite of their skeletons. ^{16}O is lighter than ^{18}O, so water molecules containing it

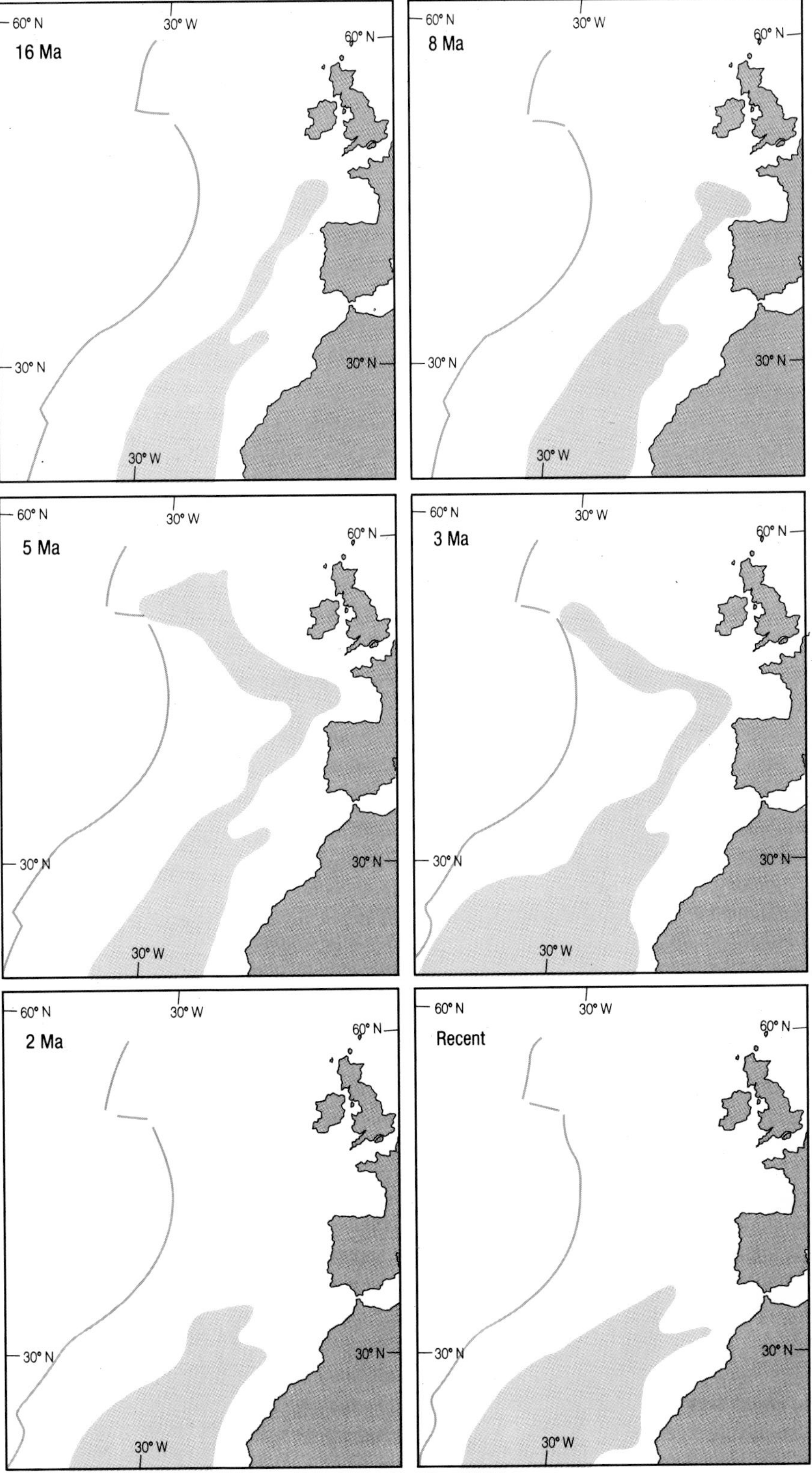

Figure 3.8 The extent of penetration of Antarctic Bottom Water (blue shaded area) into the eastern North Atlantic from the middle Miocene to the present, as inferred from the distribution of benthic foraminiferal assemblages.

($H_2{}^{16}O$) are preferentially removed by evaporation, leaving seawater relatively enriched in the heavier molecule ($H_2{}^{18}O$). As the water removed by evaporation is eventually precipitated in polar regions and becomes 'locked up' in ice-sheets, so the average ratio of ^{18}O:^{16}O in the oceans will increase.

The ratio of ^{18}O:^{16}O in calcite skeletal material varies according to that of the water in which the organisms live. It is higher when polar ice-caps are large, and lower when they are small.

Accordingly, would you expect high ^{18}O:^{16}O ratios to be correlated with high or low stands of global sea-level?

When ^{18}O:^{16}O ratios are high, the oceans have been relatively depleted of $H_2{}^{16}O$, by water frozen into ice-sheets. Water has been removed from the oceans, so sea-level should be lower. Detailed research has established a good correlation between variations of oxygen isotope ratios in (especially) foraminiferal skeletal material and sea-level fluctuations resulting from glacial oscillations of the Pleistocene.

In addition to providing information about sea-level, oxygen isotope ratios can be used together with species distributions to determine how water temperatures have changed (the ^{18}O:^{16}O ratio in calcite skeletal material of a given species is greater in cold than in warm water). In the eastern North Atlantic (Figure 3.8), for example, bottom water temperatures were about 5°C during the middle to late Miocene, falling to about 2.5°C during the Pleistocene.

Siliceous sediments can also provide palaeoceanographic information, and though they are perhaps not quite so versatile as calcareous sediments, they can be used to supplement information obtained from them. Thus, assemblages of benthic and planktonic Foraminifera and siliceous microfossils suggest that latitudinal temperature gradients were weak when the southern Atlantic was opening in Upper Cretaceous times; and that they remained so until about the middle to late Eocene, *c.*45–50Ma ago.

The Indian Ocean is particularly interesting because of the seasonal reversals of winds and currents that occur there (the monsoons). One of the more spectacular monsoonal reversals occurs in the north-west Indian Ocean, where the Somali Current flows north-eastwards for half the year (April to September), accompanied by strong upwelling and high biological productivity, and south-westwards for the other half of the year (October to March), with no upwelling. Drill cores have been obtained in this region in order to document the history of the Somali Current. Biogenic sediments rich in diatom and foraminiferal remains are good indicators of upwelling (Chapter 1), and such sediments become prominent in the record during the middle Miocene. The implication is that (seasonal) upwelling was not an important feature of the north-west Indian Ocean before about 10–12Ma ago, suggesting in turn that monsoonal reversals may have been weak or non-existent before that time.

In the upper part of the sedimentary record in this region, representing the glacial–interglacial alternations of the Pleistocene, the sediments show a strong cyclical character; they are alternately enriched and depleted in biogenic remains typical of upwelling and in land-derived pollen (brought from inland by south-westerly winds). The cycles are correlated with glacial to interglacial climatic oscillations. It is concluded that the (seasonal) upwelling was stronger during glacial epochs, weaker during interglacials—the Earth is presently enjoying an interglacial.

These oscillations are considered to have resulted mainly from changes in configuration of the Earth's orbit round the Sun, and hence to changes in the intensity of solar radiation at the Earth's surface. A possible contributory factor was the effect on atmospheric circulation patterns of the uplift of the Himalayan mountains, which is believed to have become more rapid in the past 3Ma. By influencing the winds in the upper atmosphere, the rise of this great barrier acted to intensify the seasonal contrasts between the (winter) high pressure and (summer) low pressure systems over Asia; these contrasts were greater during glacial than interglacial epochs, and thus led to enhanced seasonal upwelling in the north-west Indian Ocean.

3.4 SUMMARY OF CHAPTER 3

1 Biogenic sediments are most abundant where productivity in surface waters is high and terrigenous sediments are scarce. The geographical separation of carbonate and siliceous sediments is related to the preservation potential of the planktonic organisms and the chemistry of the water column. Biological aggregation is important in the transport of planktonic skeletal remains to the sea-bed.

2 Carbon dioxide gas dissolves in seawater to form carbonic acid and its dissociation products. These ions are also supplied to the oceans by rivers. Total dissolved inorganic carbon (ΣCO_2) increases in deep water chiefly because of the decomposition of organic matter in the water column (liberating CO_2 which goes into solution). The greater the value of $[\Sigma CO_2]$, the more acid the water, the more likely it is to dissolve calcium carbonate.

3 The solubility of calcium carbonate increases with depth in the oceans. Surface waters are supersaturated with respect to calcium carbonate, and deep waters are undersaturated. The degree of saturation can be determined from measurements of ΣCO_2 and alkalinity, A, from which concentrations of CO_3^{2-} in the water can be calculated and compared with the concentration appropriate to $CaCO_3$ saturation.

4 The depth at which dissolution of carbonate is observed to commence is called the lysocline. It lies at or below the depth where the water is shown from measurements to be saturated with respect to carbonate. The carbonate compensation depth (CCD) is defined as the depth at which the carbonate content (usually calcite content) of sediments is 20% or less. It tends to be depressed (deeper) beneath areas of high biological productivity in the open oceans. Below the lysocline, the rate of carbonate dissolution is controlled by the value of $-\Delta CO_3^{2-}$, the difference between the saturation $[CO_3^{2-}]$ and the actual $[CO_3^{2-}]$ in the water column. Both lysocline and CCD are shallower for aragonite than for calcite.

5 Direct solution of atmospheric CO_2 makes seawater more acid, because negative charges on the bicarbonate and carbonate ions so produced are balanced by hydrogen ions (H^+). This process tends to

make the levels of the lysocline and CCD shallower. Bicarbonate and carbonate ions entering the ocean from rivers (and produced by weathering of rocks on land) are balanced by cations other than H^+ (i.e. Ca^{2+}, Na^+, K^+, etc.), and in general do not make seawater more acid (it is sometimes said that rivers add alkalinity to seawater). This process will in general not affect the levels of the lysocline and CCD.

6 The global level of the CCD was shallower in Cretaceous times than it is now, possibly: (a) because marine transgressions led to increased biological production and carbonate sedimentation on broad continental shelves; and (b) because widespread anoxia in ocean waters led to increased acidity.

7 Seawater at all depths is undersaturated with respect to silica, and the preservation of siliceous sediments depends upon the survival of siliceous skeletal debris as it descends the water column. The likelihood of preservation is greater once the remains have survived descent through surface waters. The chances of siliceous sediments accumulating are greatest where surface productivity is high, and where water depths are great, so that dilution by terrigenous or calcareous material is low. Siliceous sediments are most abundant at high latitudes in the Pacific Ocean, in the equatorial regions of both the Pacific and Indian Oceans, and in coastal upwelling areas.

8 Different marine organisms inhabit different water masses. Assuming that their ecological requirements do not change significantly over geological time, the remains of such organisms in biogenic sediments can be used to determine past distributions of water masses and patterns of ocean circulation—as well as fluctuations of the CCD.

Now try the following questions to consolidate your understanding of this Chapter.

QUESTION 3.5 Locate the Ninety-East Ridge in the Indian Ocean on Figure 1.5 (as the name implies, the ridge is near longitude 90° E and is aligned north–south). Now look back to Figure 1.7. From the nature of the sediments recovered in that core from the ridge, would you infer that the ridge has always been a relatively upstanding feature, or that it has only recently become one?

QUESTION 3.6 Explain briefly why neither calcite nor aragonite is likely to be accumulating to any significant extent on the sea-bed at the station represented by the profile in Figure 3.3.

QUESTION 3.7 Give two reasons why CO_2 will be released to the atmosphere if deep water upwells to the surface in tropical latitudes.

CHAPTER 4 THE SUPPLY OF TERRIGENOUS SEDIMENTS TO THE DEEP SEA

During continental weathering, rocks are fragmented into progressively smaller pieces by physical weathering; their minerals are decomposed and reconstituted into new minerals (chiefly clay minerals) by chemical weathering, which also releases ions into solution. Breakdown by physical weathering is the dominant process in cold, high latitude regions, whereas chemical weathering is dominant in the warm, wet low latitudes. During the process of transport to the oceans, physical breakdown continues and, if the sediment is water-borne, so also does chemical attack. At the same time, coarse sediment is progressively sorted from fine sediment and deposited in the continental areas, or in shallow coastal water. In general, except in polar regions where ice transport is important, only grains finer than gravel (less than 2mm diameter) reach the outer **continental shelf** and, eventually, the ocean basins.

This does not mean to say that only sand, silt and clay-sized sediments actually occur on the submerged continental shelves. During the Quaternary glacial periods, when water was withdrawn from the ocean basins as land ice, sea-levels were considerably lower than they are at present.

What effect must this have had on the supply and nature of sediment deposited on what are now the continental shelves?

Much of what is now continental shelf was exposed above sea-level as continental coastal plains, leaving only a very narrow continental shelf below sea-level. Rivers and glaciers deposited fluvial and glacial sands and gravels on what is now submarine continental shelf, not far from the shelf edge. Consequently, present-day continental shelves have considerable **relict** deposits of coarse sediment which were not originally deposited in a marine environment.

By the time terrigenous sediment reaches the deep ocean basins, physical and chemical breakdown are usually well advanced. The particles consist mostly of quartz grains (hard, and resistant to chemical weathering) and clay minerals (the decomposition products of minerals such as feldspar and mica).

QUESTION 4.1 Look back at Figure 1.4. Which sort of land-derived sediment represented on the map is likely to contain coarse material showing least evidence of chemical breakdown?

Glacial sediments are latitudinally localized in their occurrence and contribute about 10% of the sedimentary materials that reach the oceans. The largest input of sediments comes from rivers which provide more than 80% of the total. Over two-thirds of the sediment supplied by rivers to the oceans is derived from the great rivers of southern Asia and from the larger islands in the Pacific and Indian Oceans.

By no means all of the sediment brought in by rivers reaches the deep oceans. Round the Pacific margins, much is trapped in **marginal basins** behind island arcs, or is deposited in trenches. Round the Atlantic and Indian Ocean margins, sediment is deposited on the continental shelf where it is reworked by waves and currents until eventually some of it escapes over the shelf edge into the deep ocean basins.

4.1 PROCESSES AT THE SHELF EDGE

Most of the sediments forming the sea-bed along the shelf edge are relict sediments, deposited during the last glaciation, when sea-level was lower. They are generally not covered by modern muds, and they contain little fine-grained material. The proportion of mud appears to increase on the upper parts of the **continental slope**, which suggests that it has been winnowed out of the reworked shelf-edge sediments and redeposited lower down, below the shelf break. Currents of sufficient strength must occur in the waters over the shelf edge frequently enough to prevent the build up of any mud that is being deposited there.

4.1.1 WATER MOVEMENTS AT THE SHELF EDGE

There appear to be four main ways in which currents might develop at the sea-bed in the vicinity of the shelf break.

1 They can result from the large, wind-generated, surface waves that occur during severe storms—but these are relatively rare events and the associated bottom currents would be active at the shelf edge for only a few hours each year, on average. Symmetrical ripples found on sediment surfaces at a depth of 200m off Oregon (western USA) suggest that, in the Pacific at least, storm waves can move sand at these depths.

2 In areas where the crests of tidal waves are oriented parallel to the shelf edge, the associated tidal currents will be at right angles to it, and bottom currents may reach their maximum speeds at the shelf break.

3 The decrease in atmospheric pressure associated with severe storms can lead to significant elevations of sea-level, and these elevations move with the storms as long waves. Where a storm system is travelling off the shelf, the associated long wave may generate bottom currents that are strongest where the water shallows along the shelf break.

4 Experimental and theoretical data also indicate that **internal waves** (which develop along density discontinuities called **pycnoclines**) can cause movement of sediments by breaking on the continental slope.

All of these processes are likely to give rise to to-and-fro movements of water that are strong enough to resuspend fine material, which then escapes from the shelf edge, down the slope.

4.1.2 THE ESCAPE OF SUSPENDED MATERIAL

There are two important mechanisms by which transport of suspended material at the bed can occur.

1 **Lutite flows** This is the name given to low-density turbidity currents in which concentrations of suspended material do not exceed a few tens of milligrams per litre. (The term lutite simply refers to material made up mainly of clay-sized and silt-sized particles.) The excess density imparted to the water by these low concentrations is not enough to drive a current through the density stratifications of the full open ocean water column

and down to the deep sea-floor. In any case, on the low gradients of most open slopes, such currents are too slow to move far; they swing round parallel to the slope because of the **Coriolis force** and there is a gradual deposition of the suspended material. This is presumably the origin of the fine sediments found on the upper parts of continental slopes.

2 **Cascading** Density stratification can develop in mid-latitude shelf seas during the summer months. This is broken down in winter by the combined action of winds and tidal currents, and the whole water column can be cooled and mixed, provided the sea is shallow enough. When this happens near the shelf edge, the shelf water may become colder and denser than the adjacent slope water, and will cascade off the shelf and down the slope, carrying suspended sediment with it. If cascading occurs during a stormy season it can be an effective mechanism for carrying away resuspended sediment, but it is unlikely to extend far down the continental slope because the deeper layers of the ocean are usually denser than the cascading water.

Do any of the processes we have considered so far suggest a way of transporting large amounts of sediment into the ocean basins?

The short answer is no. Material carried directly over the shelf edge does not travel further than the upper continental slope, either in lutite flows or cascading water. In canyons, both lutite flows and cascading can increase the movement of suspended material that is already moving downwards; but by themselves these processes do not seem capable of moving large amounts of sediment from shelves to ocean basins, even allowing for increased transport during storms.

4.2 TURBIDITY CURRENTS AND OTHER GRAVITY FLOWS

Much of the material that accumulates on the upper continental slope is in an unstable situation and likely to move down the slope. Such movements can be classified according to the degree of internal deformation of the mass of sediment that is moved. They range from slides and slumps (in which deformation is minimal) through debris flows (with moderate deformation) to turbidity currents (in which the sediment is dispersed as a turbulent suspension in seawater). Slides and slumps can develop into debris flows and turbidity currents, depending on a variety of factors, such as the volume and nature of the sediments involved, their degree of compaction and water content, the angle of inclination (dip) of sedimentary layering, the gradient of the continental slope, the intensity of the triggering mechanism, and so on.

4.2.1 SLIDES, SLUMPS AND DEBRIS FLOWS

Slides and slumps result from mechanical failure along inclined planes allowing large masses of sediment to slip (in a manner very similar to landslips). These masses may be tens to hundreds of metres thick and hundreds to thousands of metres long and wide. There is relatively little deformation of the sediment—so that original layering can still be detected—but the movement gives rise to a bulging hummocky topography at the base of the continental slope and on the upper part of the continental rise. Scars are left higher up on the continental slope

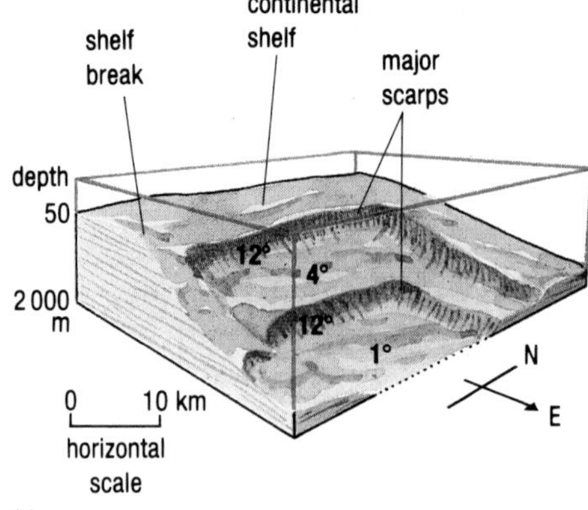

(a)

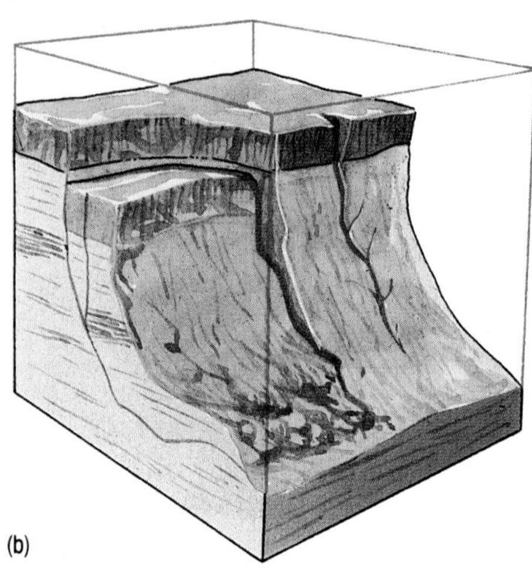

(b)

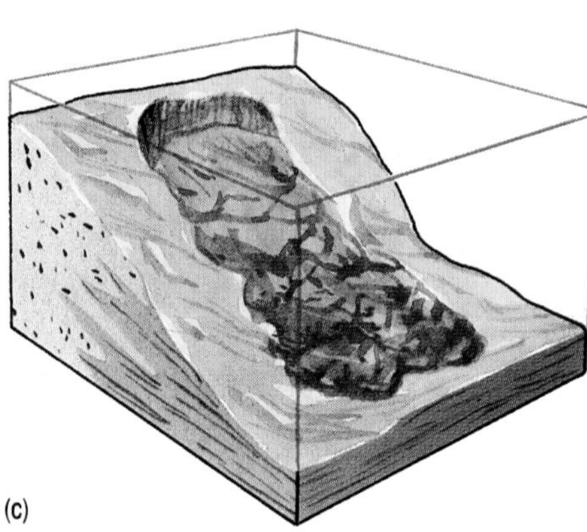

(c)

Figure 4.1 (a) General schematic view of the late Pleistocene Currituck slides, north-east of Cape Hatteras (36°30′N, 74°30′W). The smaller slide occurred after the larger one, and the total volume of sediments involved was about 130km^3. In this case, failure occurred along bedding planes between successive layers of sediment, which are tilted seawards at about 3°. In contrast, slumps (b) occur along curved failure surfaces that cut across bedding planes, and the slumped sediments retain some coherence, except at the base of the slump. However, in debris flows (c), all coherence of original sedimentary layering is lost.

where the slides and slumps originated—analogous features can be observed above many landslips (Figure 4.1(a) and (b)). Failure may occur on slopes of as little as 2° and may be triggered by earthquakes, or may simply occur after the relatively rapid accumulation of sediments that are unstable because of their high water content. The distinction between submarine slides and slumps is not always obvious (nor is it usually important) and you may find the terms used interchangeably in the literature.

One of the world's largest known slides is found in the Storegga area off the west coast of Norway (about 62°N). Three slide events are recognized; the first occurred between about 30000 and 50000 years ago, the other two some 6000 to 8000 years ago. The slides also triggered debris flows and turbidity currents, and transported a total of nearly 6000km^3 of sediments from the continental shelf break down to water depths as great as 3500m and over distances up to 800km. They are believed to have been set off by a combination of earthquake activity and the decomposition of gas hydrates in the sediments (see Section 5.3.3).

Debris flows (Figure 4.1(c)) originate for the same reasons as slides and slumps, and they involve the sluggish movement of a mixture of sediments down the slope. Internal deformation is sufficient to obliterate all traces of original layering, but the material retains some coherence. Debris flows typically comprise material that ranges from boulders to clay particles, and they can flow for considerable distances over slopes of as little as 0.1°. They are dominated by the cohesive nature of the clay matrix, the strength of which is mainly responsible for supporting the larger particles. Several examples of large debris flows have been discovered along continental margins. The flows shown off north-west Africa in Figure 4.2, for example, involved the displacement of about 600km^3 of sediment; this is as much terrigenous sediment as the nearshore basins off southern California have received from turbidity currents in a million years.

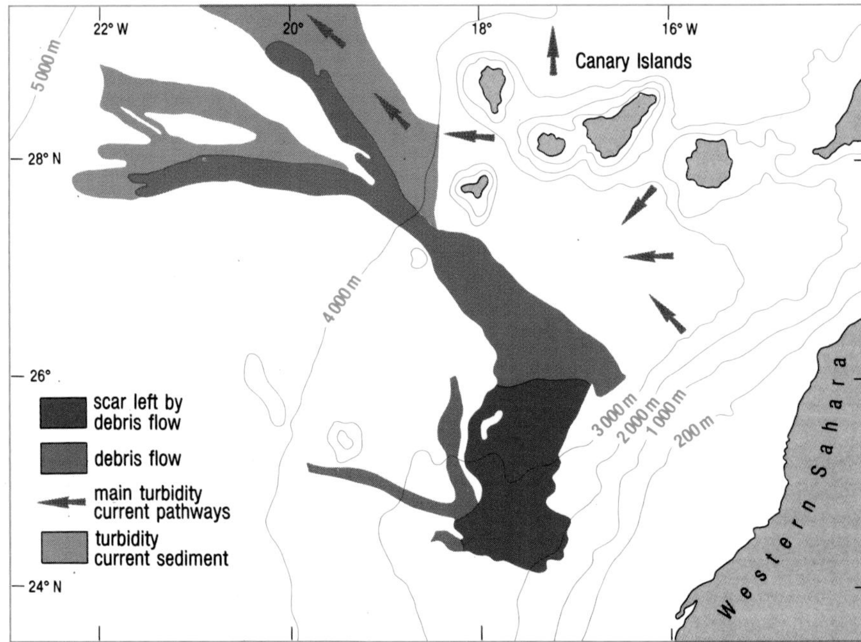

Figure 4.2 The extent of a debris flow off north-west Africa, and main pathways of associated turbidity currents.

Slides, slumps and debris flows in general contribute to development of the **continental rise** at the base of the continental slope. Transport of sediment out onto the abyssal plains is mainly by turbidity currents.

4.2.2 TURBIDITY CURRENTS

Turbidity currents are high-velocity density currents, which generally carry larger amounts of suspended sediment than lutite flows and are therefore denser. They are considered to be responsible for incising **submarine canyons**, for the breakage of submarine cables, and for the formation of submarine fans and the deposition of sand layers on **abyssal plains**. Before looking at the deposits they form, we shall examine the nature of the turbidity currents themselves, for they have some very interesting features.

Turbidity currents are probably triggered by the same mechanisms that set off slides, slumps and debris flows, and they are formidable transporters of sediment from the continental slope to the deep oceans. These huge masses of sediment-laden water can travel at up to 90 km h^{-1} (25 m s^{-1}), carrying up to 300 kg m^{-3} (300 g l^{-1}) of material in suspension, including gravel-sized particles as well as large quantities of mud, and they can transport this material up to 1000 km from the source. Their flow is not necessarily confined to canyons or channels, but may advance over a broad front, only parts of the flow being channelled. Canyons and channels may, however, give some idea of the heights of turbidity currents, particularly when such canyons and channels have levées formed of the material accumulated along the banks when the currents overflow them. Levées have been seen on canyons as deep as 1000 m below the surrounding sea-bed, but flows big enough to fill these must be very rare. Channels on deep-sea fans on the upper continental rise are often 100 m deep and they can reach a depth of 300 m. As the currents emerge from the channels and canyons on to a relatively level plain, their height probably decreases rapidly, to a few tens of metres. At least some of the larger submarine canyons in which turbidity currents flow were incised into the continental slope during the Quaternary periods of low sea-level when rivers flowed across what are now continental shelves, and disgorged their water and sediment loads very close to the shelf edge.

Turbidity currents that develop from initial slides or slumps of the kind described in Section 4.2.1 are less likely to be associated with submarine canyons, because they can occur almost anywhere on an unstable continental shelf edge. Canyon systems are found off the mouths of major rivers, where there is a continuing supply of sediments to a particular part of the shelf, so that failure occurs repeatedly in the same general area. Turbidity currents generated off the mouths of rivers entering deep fjords almost always form canyons, though these tend to be rather small-scale features, compared to those incised into more open continental slopes.

The upper parts of canyon systems can consist of a number of smaller tributaries that coalesce into a single major canyon lower down on the continental slope (e.g. the Ascension canyon system off California). Channels have meanders in much the same way as rivers do on land (Figure 4.3(a)), and they can be abandoned and infilled as new ones are cut (Figure 4.3(b)).

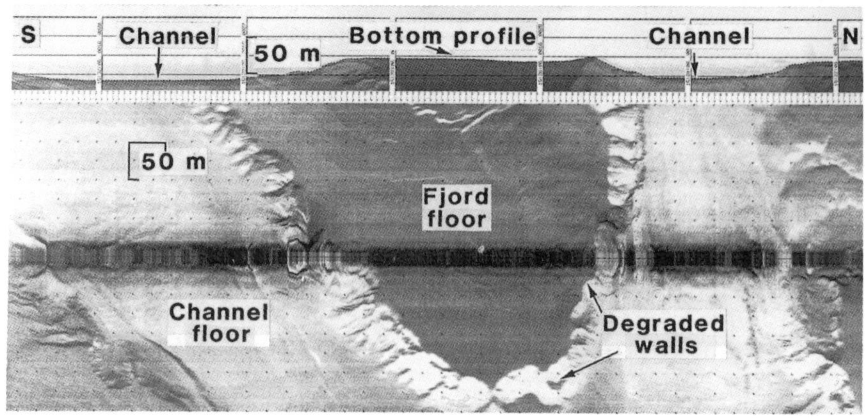

Figure 4.3 (a) Sidescan sonar pictures (plan view and section) of a bend in the incised channel formed by turbidity currents flowing along the floor of a fjord in British Columbia.

(b) Interpreted seismic reflection profile of part of the California continental slope, showing an abandoned and infilled channel of the Ascension submarine canyon system which was oblique to the continental slope. Note the greatly exaggerated vertical scale.

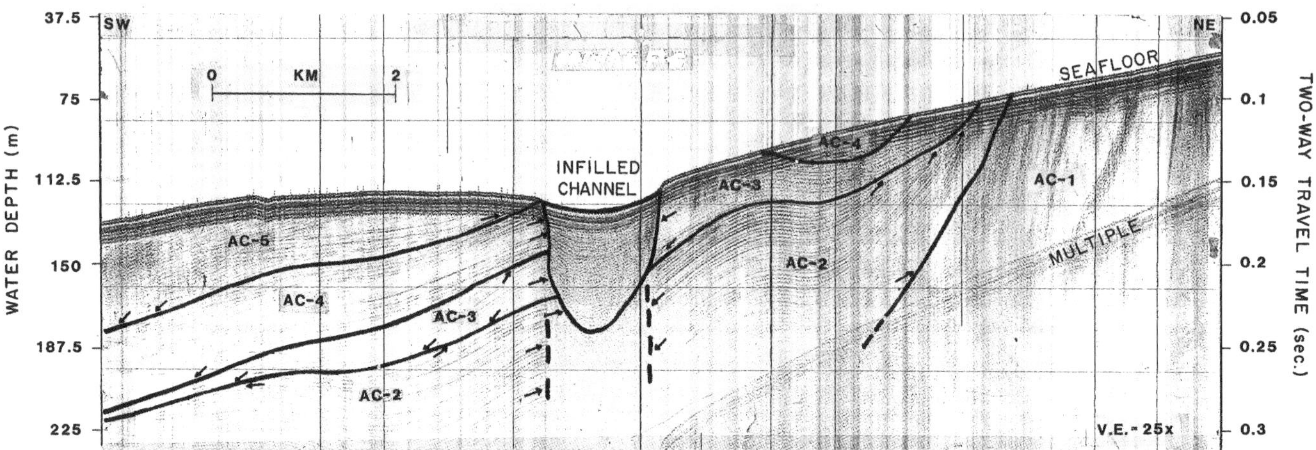

Important indirect evidence about the speeds of turbidity currents has been obtained from studying the breaks in submarine cables crossed by the currents. The most often-quoted example is the turbidity current which resulted from an earthquake on the Grand Banks, Newfoundland, in November 1929 (Figure 4.4). Following the earthquake, 12 submarine telegraph cables were broken in at least 23 places over a period of about 12 hours. At first, it was assumed that the earthquake had caused the damage by itself, and it was not until 1952 that the breaks were related to a turbidity current. It appeared that the earthquake had triggered a massive slump on the continental slope, and this broke submarine cables in the immediate vicinity. A turbidity current deriving from the slump travelled on down the slope, partly along shallow channels but mainly as a broad-fronted flow.

It is believed that this particular flow extended about 800km from its source, across the abyssal plain, before it stopped, and that at its fastest it reached between 40 and 55km h^{-1} (between about 11 and 15m s^{-1}). There is evidence that many turbidity currents travel even faster, commonly attaining maximum speeds of around 25m s^{-1} (*c.* 50 knots, or 90km h^{-1}), slowing to less than 10m s^{-1} (*c.* 20 knots, or 36km h^{-1}) at the extremities of submarine fans and to only 0.1–0.2m s^{-1} (<0.5 knot, 1km h^{-1}) on the outer parts of abyssal plains.

The densities of turbidity currents have not been directly measured either, and estimates cover an appreciable range of values. Densities as

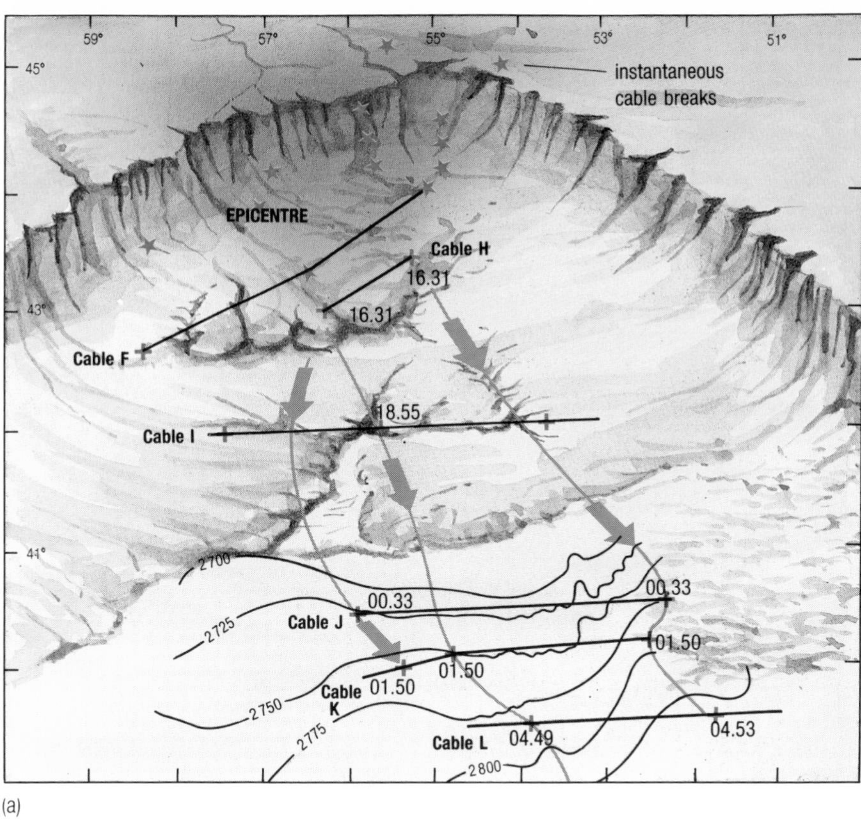

Figure 4.4 (a) Combined map and perspective view of the area of the Grand Banks earthquake in 1929, showing the epicentre, the cables broken by the turbidity current(s) and the times at which they were broken. The first cables were broken an hour after the earthquake, which occurred at about 15.30 hours. The sea-bed lies at about 2000m depth in the vicinity of cables F and H. Stars in the red shaded area show cable breaks due to the earthquake itself (see (b)), and crosses show breaks due to the turbidity current afterwards.

(b) Interpreted seismic reflection profile of the Grand Banks area, showing the configuration of the initial slump in relation to the earthquake epicentre and the first cable break by the turbidity current. Stars and crosses as in (a).

high as 1.5×10^3 to $2 \times 10^3\,\mathrm{kg\,m^{-3}}$ have been proposed, but these are more likely to be appropriate to mud flows. For a predominantly sand-sized load, densities would probably not be much more than $1.15 \times 10^3\,\mathrm{kg\,m^{-3}}$, but mud-laden water with densities of up to $1.4 \times 10^3\,\mathrm{kg\,m^{-3}}$ could be mixed with sand and still give a viable turbidity current. It seems likely that densities mostly lie in the range 1.03×10^3 to $1.30 \times 10^3\,\mathrm{kg\,m^{-3}}$. The lower end of this range is very close to the density of seawater, and so currents with these low densities would essentially be lutite flows (Section 4.1.2).

The dynamics of turbidity currents

A good deal of theoretical and experimental research has been carried out on turbidity currents. Experiments show that after the initial surge of sediment and water has occurred, a turbidity current divides into two main parts; a *head*, and a *body* which extends back into a tail (Figure 4.5(a)). The height, or thickness, of the head is usually at least twice that of the body. This division has some interesting aspects.

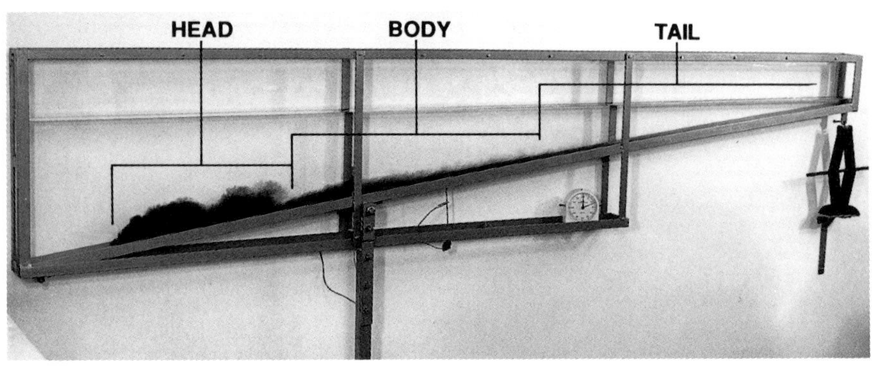

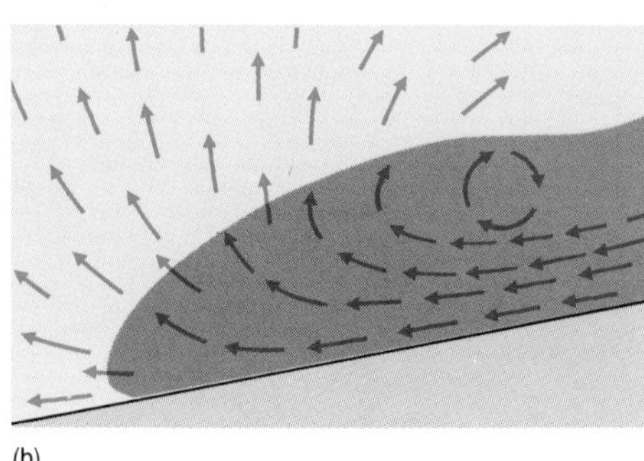

(b)

Figure 4.5 (a) The structure of a 'turbidity current' generated when dyed salt water (high density) is allowed to flow down a slope beneath normal tap water in an experimental tank.

(b) The flow of fluid in and around the head of a turbidity current.

The head and the body travel at different speeds, at least in the early stages of flow. The speed of the *head* depends mainly on the difference in density between the flow and the surrounding seawater, and on the thickness or height of the head, but *not* on the angle of slope. It is the speed of the *body* that is controlled by the angle of the slope down which the current moves (as well as by the density contrast and other factors, of course).

The speed of the head, u_1 is given by:

$$u_1 = A\sqrt{\frac{(\rho_t - \rho)}{\rho_t} g h_1} \tag{4.1}$$

where: ρ_t and ρ are the densities of the flow and the surrounding water, respectively;

g is the acceleration due to gravity;
h_1 is the height of the head; and
A is an empirical constant with the value 0.7.

Equation 4.1 looks rather complicated, but it expresses a very simple relationship, namely that the only variables determining the speed of the head of a turbidity current are its height and the density contrast between the flow and the surrounding water. Small changes in these variables can affect the speed significantly, because they are related to the square of the speed:

$$u_1^2 \propto (\rho_t - \rho)\, h_1 \tag{4.1a}$$

QUESTION 4.2 (a) Use equation 4.1 to work out an approximate height for the head of a turbidity current, given an average density of 1.20×10^3 kg m^{-3} (i.e. somewhere about the middle of the generally accepted range), and a speed of 20 m s^{-1}. Take g to be 9.8 m s^{-2}, and assume the density of seawater to be 1.0×10^3 kg m^{-3} (to keep the sums simple).

(b) Would the height of the head be greater if (i) its density were greater, or (ii) its speed were less?

If we could see it, therefore, the front of a high-density turbidity current would be an awesome sight: probably hundreds of metres high and moving at something like 70–80 km h^{-1}, at least in the early stages. What is more, the coarser sediments, including gravels in some cases, are concentrated in the head, which is therefore extremely erosive.

The next question that must arise is: if these high speeds are not related to the slope over which the current is travelling, how are they maintained? Experimental and theoretical studies show that in the earlier stages of the flow, the body is moving faster than the head. Water and sediment are forced into the head, and the excess pressure lifts the additional fluid up within the head (Figure 4.5(b)), and returns it to the body by waves breaking along the steep irregular interface at the back of the head (you can see them on Figure 4.5(a)).

As we noted above, it is the speed of the body, u_2, which in fact depends on the angle of the slope:

$$u_2 = \sqrt{\frac{\rho_t - \rho}{\rho_t} g h_2 \frac{\sin \beta}{C_f}} \tag{4.2}$$

where:

h_2 is the height of the body;

β is the angle of the slope; and

C_f is a coefficient for the combined effects of friction at boundaries between the sea-bed (the base of the flow) and surrounding seawater (top and sides of the flow). It is not a constant, but we can use a value of 6×10^{-3} for most purposes. Equation 4.2 also looks complicated, but like equation 4.1 it does express quite a simple relationship between the square of the velocity, the density contrast between the flow and its surroundings, the height of the body, and the slope. By analogy with equation 4.1, in other words, we can see that for the body of the flow:

$$u_2^2 \propto (\rho_t - \rho)\, h_2 \sin \beta \tag{4.2a}$$

Equation 4.2 is significant because you can use it to show that the speed of the body can be greater than that of the head only over the initial stages of the flow. When the turbidity current reaches the base of the continental slope, it begins to slow down and eventually the speed of the body becomes equal to, if not less than, that of the head. We can estimate a limiting angle of slope at which this might happen.

QUESTION 4.3 You worked out a height of 500 m for the head in Question 4.2(a). If the body is half that height (Figure 4.5 and related text), i.e. 250 m, what must be the angle of slope for the speed of the body to be equal to that of the head, i.e. 20 m s^{-1}? Use values for the other variables as given in Question 4.2.

The angle you worked out in Question 4.2 is very small, but it is actually greater than the angle of slope at the sea-bed in the area where the cable breaks occurred following the Grand Banks earthquake (Figure 4.4): that was only about a fifth of a degree at the upper end, even less at the lower end. The answer to Question 4.3 suggests also that when the angle of slope becomes less than about half a degree, the speed of the flow as a whole no longer depends on the slope; which helps to explain why turbidity currents travel hundreds of kilometres over the ocean floor.

There is nothing especially mysterious about the equations for the speed of turbidity current flows (equations 4.1 and 4.2). The discerning reader may already have noticed that they are simply variants of one of the basic **equations of motion** of physical oceanography, namely the equation for the vertical direction/dimension, which in its simple form is the **hydrostatic equation**, relating pressure to density ($p = \rho gh$).

There is a further interesting aspect of turbidity currents which makes them different from the normal transport of sediment by water. In ordinary water flows, such as rivers or tidal currents, the movement of water over the bed generates a **shear stress** and turbulence which lifts the sediment into suspension. In a turbidity current, however, the sediment is already in suspension and it is the increased density of the sediment/water mix that causes flow to occur. The flowing turbidity current itself naturally causes shear stress and turbulence at the bed, lifting more sediment into suspension and further intensifying the flow. This feedback effect is known as *autosuspension*, and there is an empirical relationship between the settling velocity ν, of particles in a turbidity current and the speed of the current, such that when:

$$\nu \leqslant {}^{2}\!/_{3}\, u_2 \sin \beta, \qquad (4.3)$$

the speed of the current down the slope generates sufficient turbulence to hold the particles in suspension. In this equation, ν is called the **autosuspension limit**, and the angle of slope appears in equation 4.3 because particle settling has a component of motion down the slope, which acts to increase the downward movement of the fluid. Equation 4.3 can be used to estimate speeds of flow of past turbidity currents, from analysis of particle sizes of deposited sediment and knowledge of the slopes over which they travelled—which can be obtained from bathymetric data.

Deposition from turbidity currents

Most deposition occurs near the base of the continental slope. The comparatively sudden decrease in current velocity associated with the decrease in the gradient of the sea-bed leads to rapid deposition. The sediment is built up into lobe-shaped **submarine fans** that may be tens to hundreds of kilometres across (Figure 4.6(a)), and which contribute to development of the continental rise, along with slides, slumps and debris flows (Section 4.2.1). Subsequent turbidity currents flow out over the surface of the fan carving channels in the surface and extending the deposits seawards. The coarsest sediment—gravel and coarse sand—is found in the channels on those parts of the fans closest to the foot of the continental slope (Figure 4.6(b)). However, at the outer part of the fan, where current velocity is greatly diminished, finer sands and silts are deposited. The finest clays remain in suspension until eventually they settle out in the intervals between turbidity currents (Figure 4.6(c) and (d)). It is probable that some of this fine sediment is carried out into the deepest parts of ocean basins to be deposited along with pelagic sediments.

Thus, the typical products of most turbidity currents are the successions of gravels, sands, silts and muds (clays) known as **turbidite sequences** (Figure 4.6(c)). However, it is clear from Figure 4.6(b) that different parts of a sequence will be found in different parts of the submarine fan—it is rare for the full sequence to be recovered in a single core. It is also clear from Figure 4.6(b), that, in their *terminal* stages (i.e. out on the abyssal plain), turbidity currents are typically dilute suspensions containing sediment concentrations of 1–2 kg m^{-3}. This is still a good deal more concentrated than lutite flows, but nonetheless represents a great reduction in sediment load compared with the early stages of a turbidity current flow. However, there is evidence that in some instances at least, the picture is rather different: the terminal stage of some turbidity currents may be sluggishly moving viscous suspensions, with high sediment concentrations: 50–100 kg m^{-3} or more.

This evidence is provided by turbidity current deposits of uniform (unlaminated) mud up to several metres thick and with hardly any variation of grain size (2–5 μm) throughout the full thickness, which have been recovered in cores from the Madeira abyssal plain and elsewhere (there is evidence of this in parts of Figure 4.6(d)). The mud is underlain

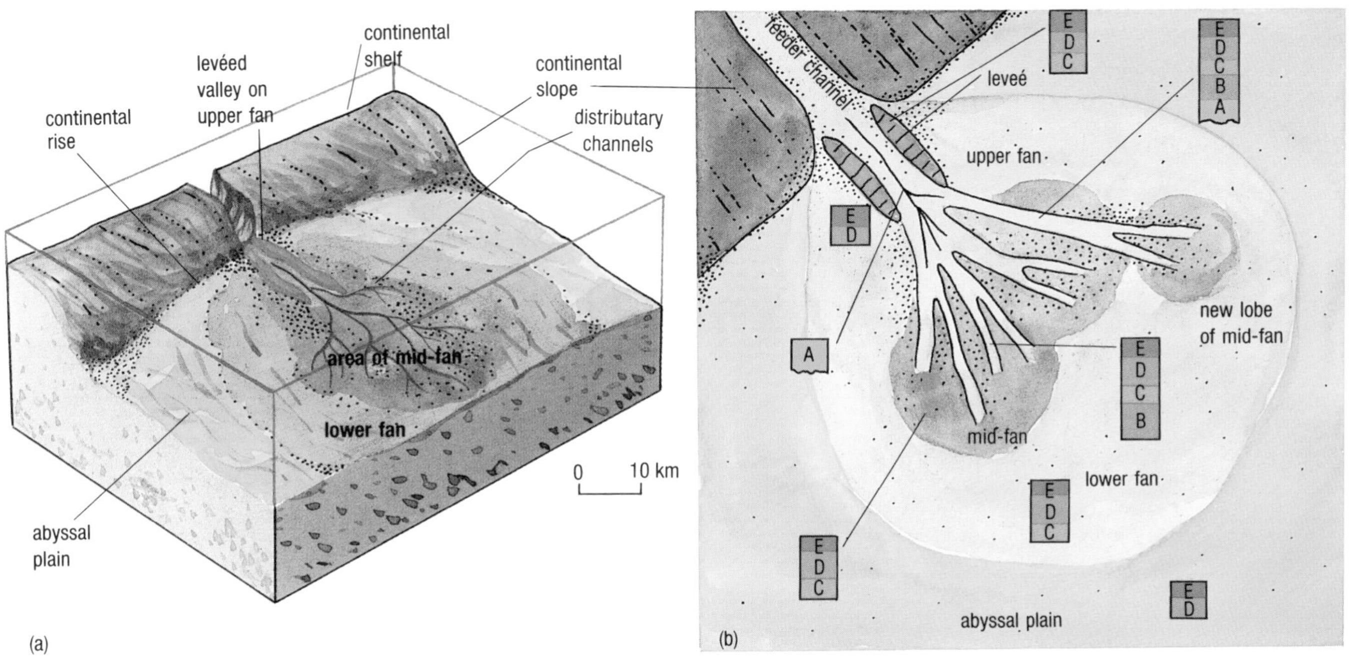

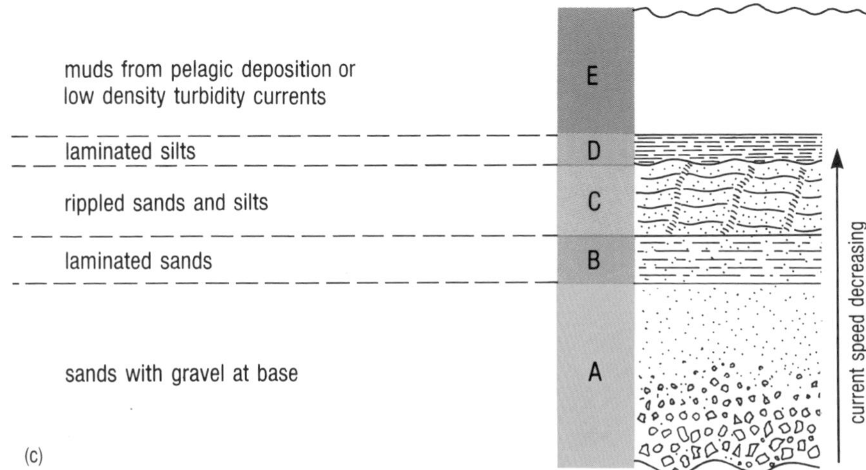

Figure 4.6 (a) General view of a submarine fan. The vertical scale is greatly exaggerated.

(b) Distribution of different types of turbidite sequences on a submarine fan. For key to letters, see (c).

(c) Schematic representation of the 'ideal' or complete turbidite sequence. The wavy line at the base of the sequence indicates an *erosion surface*, because turbidity currents erode the surfaces over which they travel as well as carving out canyons—see Figure 4.3. The upper wavy line is the base of the overlying sequence.

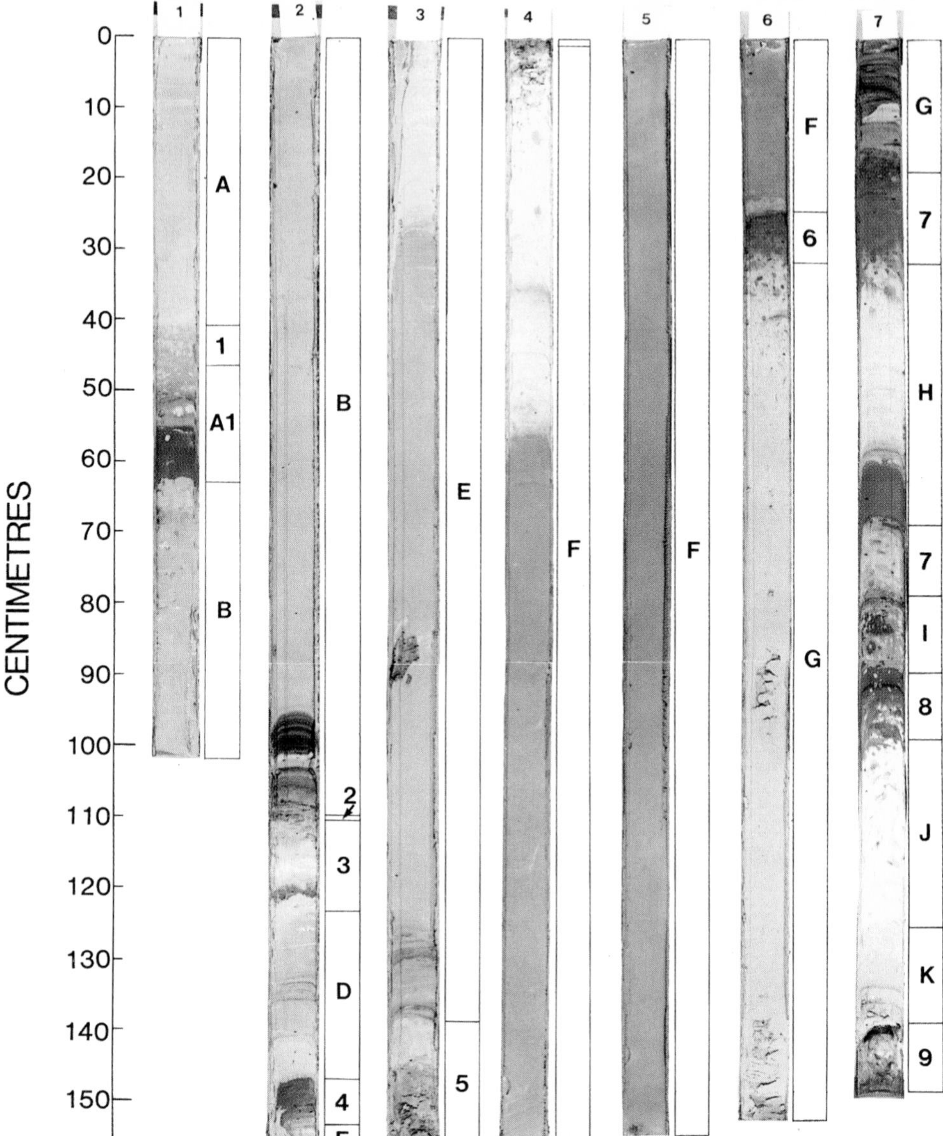

Figure 4.6 (d) Photograph of a segmented sediment core (No. D10688) from the Madeira abyssal plain. The top segment of the core (present-day sea-bed) is at upper left, the bottom segment (total length of core *c.* 10m) is at bottom right (age *c.* 275000 years). The core consists of thick layers of rapidly deposited turbidites (T) and thin sequences of much more slowly deposited pelagic sediments (P). The sediments are relatively rich in calcium carbonate, containing from 40 to 70% $CaCO_3$. Pale grey colours represent high-carbonate sediments with little organic carbon. Brown–green colours represent lower carbonate contents and moderate (greener shades) to low (browner shades) contents of organic carbon.

by layers of silt no more than a few tens of cm thick in all, in which the grain size decreases fairly regularly upwards, from about 60μm down to about 10μm or less. These silt–mud sequences are interpreted as resulting from single depositional events, and it is difficult to see how they could be deposited from dilute suspensions.

Why might that be?

The settling velocities of clay mineral particles are of the order of 10^{-6} m s^{-1}, and to deposit a layer of uniform mud several metres thick from dilute suspension would require a single flow to last for weeks or months, which seems improbable.

Instead, it is proposed that where these thick mud layers are found, the turbidity current evolved into a high density flow by particle settling within the body, as the sea-bed gradient decreased along the flow path and the current slowed down. Sediment concentrations of 50–100 kg m^{-3} could be achieved, despite the low speeds of about 0.5 m s^{-1}. The gradients on abyssal plains are extremely low. On the Madeira abyssal plain, for example, the relief is described as being less than 30 m over hundreds of km, which implies gradients of about 10^{-5} (i.e. 1 in 10^5). Under these conditions, a flow velocity of 0.5 m s^{-1} gives an autosuspension limit appropriate to particles of about 3 µm in size (*c.* 5×10^{-6} m s^{-1}). This is still sufficient to keep clay particles from settling, but it is doubtful if the autosuspension relationship (equation 4.3) can strictly be applied in these viscous suspensions.

QUESTION 4.4 Nonetheless, you can still make a first-order estimate of what the autosuspension limit actually is, using equation 4.3, the velocity data given above, and the fact that at these low gradients the tangent of the angle of the slope is the same as the sine, so you can take sin β to be 10^{-5}. What is the autosuspension limit?

Figure 4.7 illustrates the main features of such a high density mud-rich flow in its terminal stages and shows how the observed sediment distribution might be achieved. Despite the low speed, seawater continues to be entrained at the base of the head and, helped by the velocity shear at the base of the flow, creates turbulence there and so flushes clay particles up into the body, as well as contributing to the upward decrease in grain size of the basal silt layer. The mud suspension in the body of the flow is similar to the *fluid mud* that forms in some tidal estuaries. Turbulence is damped down in such suspensions, because of their cohesive and viscous nature; and in consequence water is not entrained at the upper surface of the flow to dilute the suspension. When the flow eventually stops, the mud suspension continues to dewater and is deposited as a uniform layer of muddy sediment.

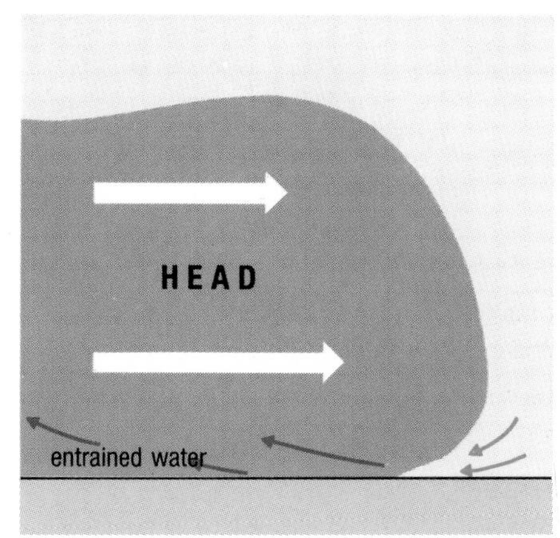

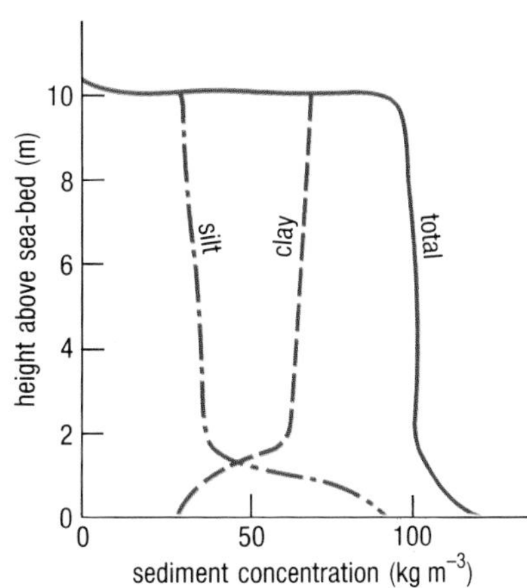

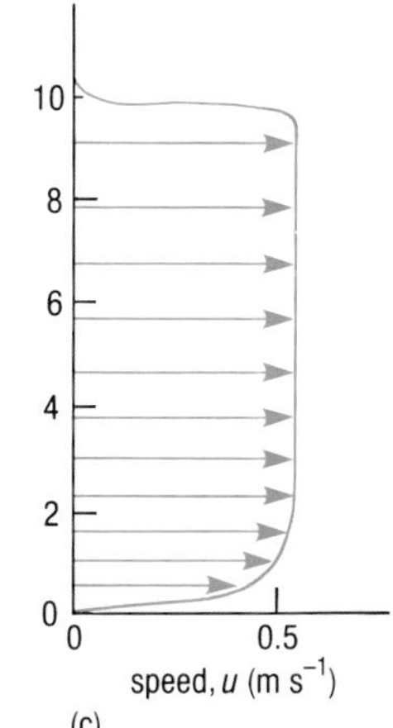

Figure 4.7 (a) Sketch of a 'fluid mud' turbidity current, showing how water entrained at the head flushes up through the body of the flow and causes some turbulence in the basal silt layer.

(b) Vertical concentration profiles showing the change in proportions of silt and clay through the body of the flow.

(c) The velocity profile of the flow.

4.2.3 TURBIDITE DEPOSITION AND ABYSSAL PLAINS

Only a rather small proportion of sediments on abyssal plains originate from 'normal' pelagic deposition. Abyssal plains are built mainly from successive turbidite deposits which have in places accumulated to thicknesses in excess of 1000m, burying the rough volcanic topography beneath (Figure 4.8). They are developed most extensively where turbidity currents are able to flow frequently and unimpeded down the continental slope and out onto the deep ocean floor.

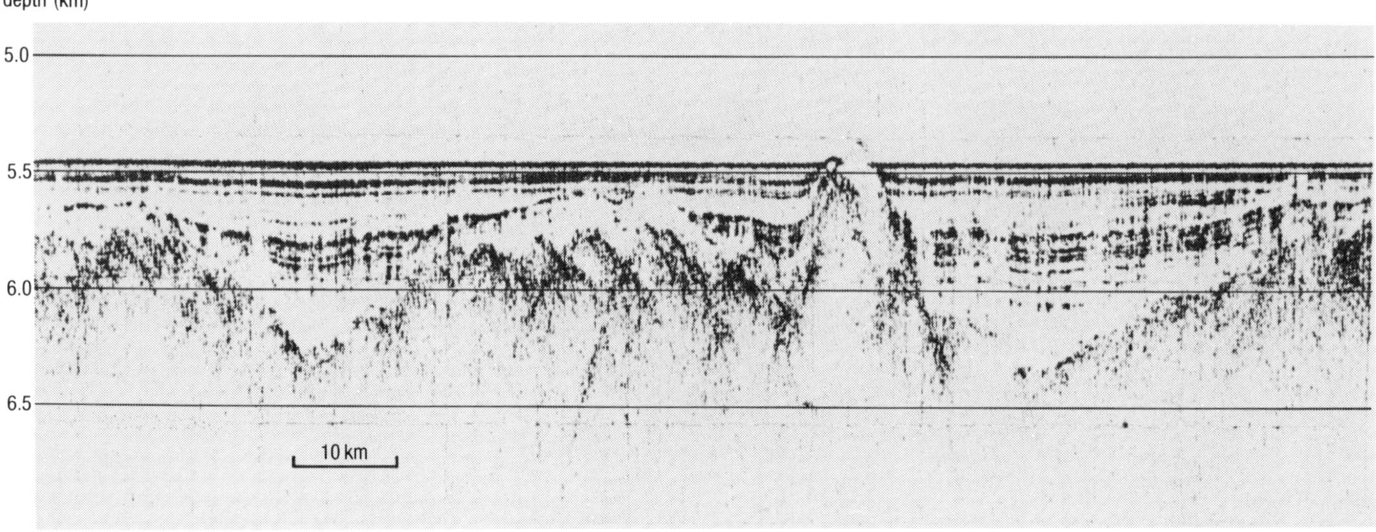

Figure 4.8 Seismic cross-section across part of the Madeira abyssal plain in the eastern Atlantic, showing rough volcanic topography buried by sediments (mainly turbidites, Figure 4.6), with only a small abyssal hill showing above the flat sediment surface.

Figure 4.9 shows the global distribution of abyssal plains formed of turbidite deposits. Note the virtual absence of these features from most of the Pacific, which is bordered by trenches and marginal basins that trap sediment and prevent turbidity currents from reaching the deep ocean floor.

During the late Cenozoic (the past 15Ma or so), there were substantial increases in rates of accumulation of sediments on abyssal plains, especially in the North Atlantic and North Pacific. These have been correlated with the onset of glaciations in the Northern Hemisphere: when sea-levels were lower, rivers flowed across broad coastal plains (formerly submerged continental shelves) and dropped much of their sediment load close to what is now the shelf edge. Influxes of fresh sediment would have been sufficient to trigger slumping and to generate turbidity currents. This is somewhat similar to the present-day situation in the northern Indian Ocean where major river systems draining the Himalayas discharge such vast quantities of sediment into the ocean that the continental slope is constantly unstable. Slumping and turbidity current activity have built up the huge Indus and Bengal submarine fans. Turbidite sequences have been found in sediment cores from the Indian Ocean as far as 2500km south of the Bengal delta region (Figure 4.9).

Evidence from the Madeira abyssal plain suggests that most major turbidity current activity is initiated mainly during periods of climatic change with rising or falling sea-levels, rather than in the 'quiet' periods between. There appears to be some correlation between the magnitude of the sea-level change and the volumes of turbidite sequences emplaced, but it is not immediately clear why this should be.

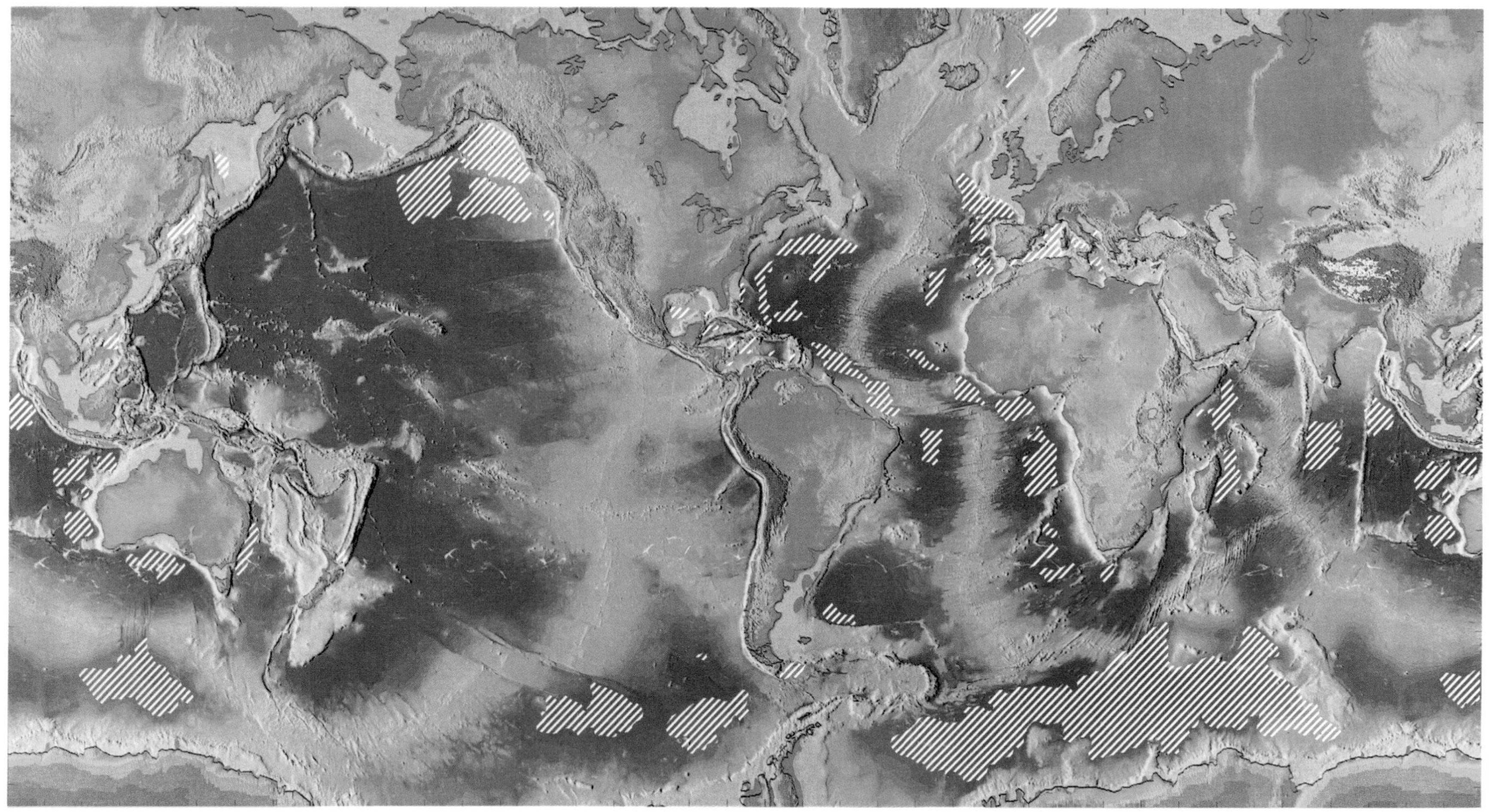

Figure 4.9 Distribution of turbidite-formed abyssal plains (diagonal lines) in the world's oceans south of about 70°N.

4.3 SUMMARY OF CHAPTER 4

1 Terrigenous sediments are the products of physical and chemical weathering at the land surface. They are transported to the continental shelves and redistributed by waves and currents. At times of low sea-level, coarse sediments were deposited on exposed shelves and these relict sediments were re-worked when sea-level rose again. Fine sediment is resuspended at the shelf edge by waves and currents. It escapes down the continental slope in low-density suspension (lutite flows) and by cascading. Where the sediments at the shelf edge are unstable, slides, slumps or debris flows occur (along with turbidity currents), carrying large quantities of coarse and fine sediment down to the continental rise.

2 Turbidity currents are the most important means whereby terrigenous sediment is transported from the continental shelf down the continental slope to the deep ocean basins. Turbidity currents may travel at several tens of kilometres per hour, and their speed is related to: the difference in density between the sediment/water mixture and the ambient seawater; the gradient of the continental slope; the height of the head and body of the flow; and frictional resistance.

3 The passage of most turbidity currents is marked by characteristic deposits known as turbidite sequences which build up submarine fans at the base of the continental slope. The terminal stages of turbidity currents may commonly be dilute suspensions of mud, but there is evidence that in some cases fluid mud suspensions develop as the flows slow down. Abyssal plains in the Atlantic and Indian Oceans are largely built from turbidite deposits but in the Pacific most deep sea sediment is pelagic because turbidites are trapped in the ocean trenches.

Now try the following questions to consolidate your understanding of this Chapter.

QUESTION 4.5 The speed of turbidity currents depends crucially on the angle of slope over which they flow. True or false, and why?

QUESTION 4.6 (a) Would you expect submarine fans to be better developed along the margins of the Atlantic or the Pacific Ocean?

(b) The rise of the Himalayas is believed to have accelerated in the past 3Ma (Section 3.3). Would you expect the number of turbidites on the Indus and Bengal fans to have increased or decreased over this period?

QUESTION 4.7 Most sediment deposition on the abyssal plains comes from normal pelagic sedimentation. True or false?

CHAPTER 5 POST-DEPOSITIONAL PROCESSES IN DEEP-SEA SEDIMENTS

'Below the thunders of the upper deep;
Far, far beneath in the abysmal sea,
His ancient, dreamless, uninvaded sleep,
The Kraken sleepeth...'

Alfred, Lord Tennyson

It is tempting to imagine that once sediment has been laid to rest on the deep sea-floor beneath several kilometres of water, far removed from waves and storms, it will remain there quite passively until such time as the ocean basin ceases to exist. Indeed, that was the consensus view until marine scientists began to take photographs of the deep sea-bed.

Sediments certainly do not lie undisturbed on the sea-bed after they are deposited. Figure 5.1 shows that even where no animals are visible, and even in the most apparently featureless parts of the deep-sea abyssal plains, there is plenty of activity. As the technology of submersibles improves and they become more widely used, more and more of the animals that make these various marks will be photographed in action. Nor is activity confined to the sea-bed itself. The mottled ooze in Figure 1.3 (b) shows clear evidence of disturbance and disruption of layers by burrowing animals (*bioturbation*).

Figure 5.1 These photographs were taken by the late Bruce Heezen, an American oceanographer, and his colleagues. The field of each photograph is about 2m across.

(a) Deep-sea red clays are found throughout the oceans at depths of 4000–4500m or greater, and cover nearly half of the Earth's surface. Vast areas of the deep ocean floor are almost featureless and such detail as can be seen is difficult to interpret. Some of the small mounds in this picture are burrows and some are faeces.

(b) A track in the centre may have been made by an acorn worm. Superimposed on this, from top left to bottom right, is a four-row trail. This was made by a holothurian, an echinoderm of the genus *Psychropotes*, see (c).

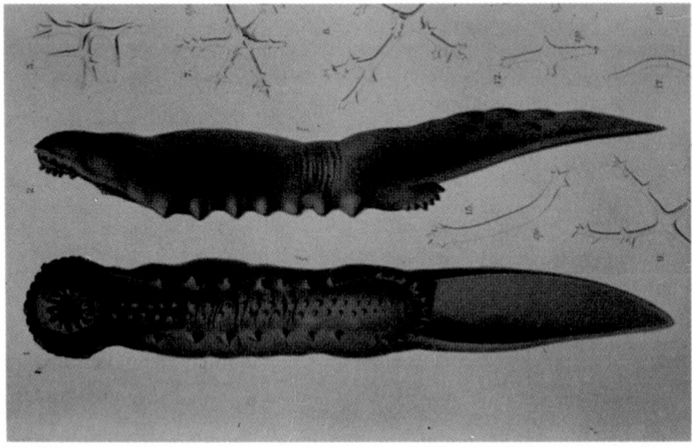

Figure 5.1 (c) A nineteenth-century lithograph of the holothurian *Psychropotes*, dredged from the Pacific. On its underside there are four rows of tube feet, corresponding to the track in (b). The tentacles around the mouth, on the left side of the picture, pick up organic matter from the sea-bed. The animals also ingest a lot of sediment which, after their gut becomes filled, is ejected onto the sea-bed in a neatly coiled pile. The lithograph also shows calcareous plates from the animal's body wall.

(d) This picture was taken at a depth of 2132m in the equatorial Atlantic. The animal in the centre is a brittle-star of the Echinodermata (Ophiuroidea), which moves over the bottom by 'rowing'—it points one 'arm' forward and pushes itself along using the other four arms. In the upper right is the remnant of a large trail, but the animal that made it was not identified.

(e) This picture was taken at a depth of 2780m in the equatorial Pacific. It is an *oblique* photograph of the sea-floor: such pictures sometimes show features more clearly than the normal vertical photographs. To the right is a deep U-shaped groove with lateral ridges, and there is also a pattern of ridges running out from each side of the groove. This may be made by a large deep-sea echinoid (sea-urchin), or a crustacean.

Biological activity is not the only cause of sediment disturbance on the deep sea-floor. Evidence of currents in the deep sea is provided not only by fortuitous glimpses of animals reacting to them, but also by bed forms, especially ripple structures in the sediments (Figure 5.2). In other words, in some parts of the deep ocean, currents are strong enough to stir up the sediment.

The different speeds of bottom currents lead to the development of a variety of bed forms on the ocean floor, generally similar to those found in shallow water—the nature of the bed forms depends mainly on the speed of flow and on the grain size of the sediments involved. For example, where the flow of a bottom current is restricted by a narrow passage so that its speed increases (as in the Drake Passage between South America and Antarctica), the bottom is scoured out and coarse residual deposits are found, sometimes in the form of ripples or ridges.

Figure 5.2 These pictures were also taken by Bruce Heezen and his colleagues.

(a) The white objects are sea-fans (related to corals, Anthozoa) growing attached to the sea-bed. They are orientated perpendicular to the flow of the current so that they can obtain the maximum food. The fish (related to the angler fish) indicates in which of the two possible directions the current is flowing: the fish points into the current, and may also depend on it to bring food. There are also brown brittle-stars entwined in the fronds of the sea-fans, and there are smaller, white brittle-stars on the sea-bed. The sea-bed has rock outcrops, running approximately from left to right. The current that feeds the sea-fans has eroded the soft sediment from the rock beneath.

(b) Ripple marks in sediments on the top of the guyot in the western equatorial Pacific, at a depth of 1000m. There are small mangangese nodules in the troughs of the ripples (see Section 5.2.1). The current is flowing south to north (there is a compass in the bottom left-hand corner).

5.1 THE NEPHELOID LAYER

The bottom few hundred metres of the water column are appreciably more turbid with sediment than the overlying seawater, because of lutite flows (Section 4.1.2) and turbidity currents (Section 4.2.2). However, the bottom waters are most turbid where there are strong bottom currents in deep water off continental margins. The turbid layer is known as the **nepheloid layer** and was first recognized using instruments called nephelometers, which record an increase in light-scattering by suspended particles. In many cases, the scattering increases by a factor of 10, going from clear water to the bottom. When the nephelometer measurements are calibrated by gravimetric methods of measuring suspended sediment concentrations, the distribution of suspended material in bottom waters can be mapped (Figure 5.3).

QUESTION 5.1 Can you offer an explanation of the distribution of suspended sediment concentrations in Figure 5.3?

The occurrence of strong bottom currents in the deep circulation is only part of the answer to this question. The high concentrations occur beneath the strong western boundary currents of the sub-tropical gyres (e.g. the Gulf Stream of the North Atlantic, or the Brazil Current of the South Atlantic), and there is evidence that the influence of these surface currents can extend even to the deep sea-bed.

The localized occurrence of very high concentrations of suspended sediment suggests there may be specific sites where resuspension takes

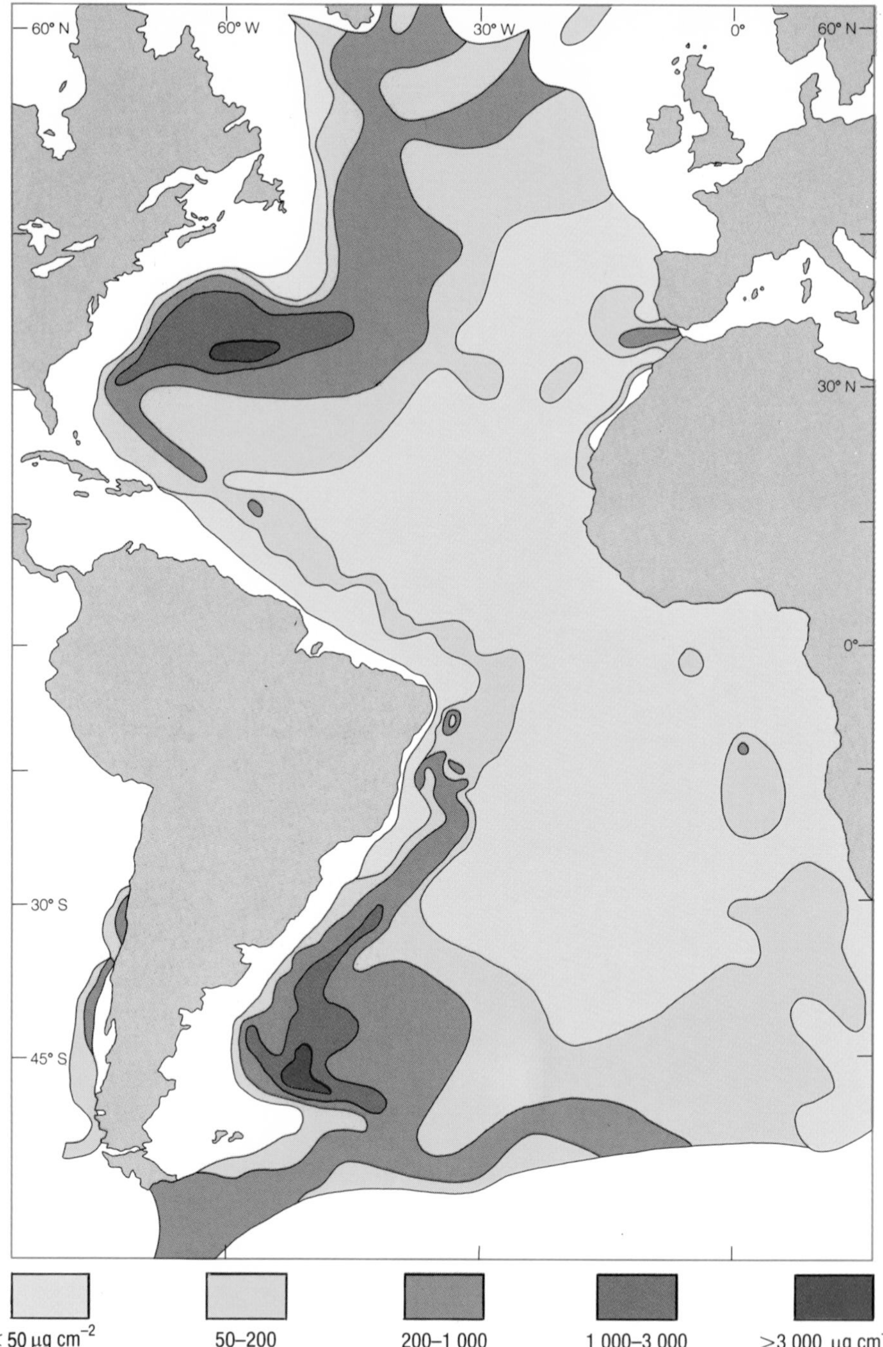

Figure 5.3 Suspended particle load in the bottom nepheloid layer of the Atlantic. The map shows concentrations of suspended matter in excess of those found in the overlying clearer water at intermediate depths.

place. These sites appear to coincide with regions where **mesoscale eddies** develop and surface current speeds are both high and variable (Figure 5.4). This surface variability and associated turbulence appear to extend to considerable depths, leading to turbulent conditions at the sea-bed. Erosive episodes, which have been referred to as **abyssal storms**, are intermittent and last for periods of days to weeks, occurring when high bottom eddy kinetic energy reinforces a strong bottom current giving overall speeds sometimes in excess of $0.3\,\mathrm{m\,s^{-1}}$, sufficient to resuspend unconsolidated fine-grained sediment, and to generate ripple marks and other features on the sea-bed (Figure 5.5).

Figure 5.4 Areas of a high surface current variability and turbulence with, superimposed, the regions of highest suspended particulate load in bottom waters. *KE*=kinetic energy.

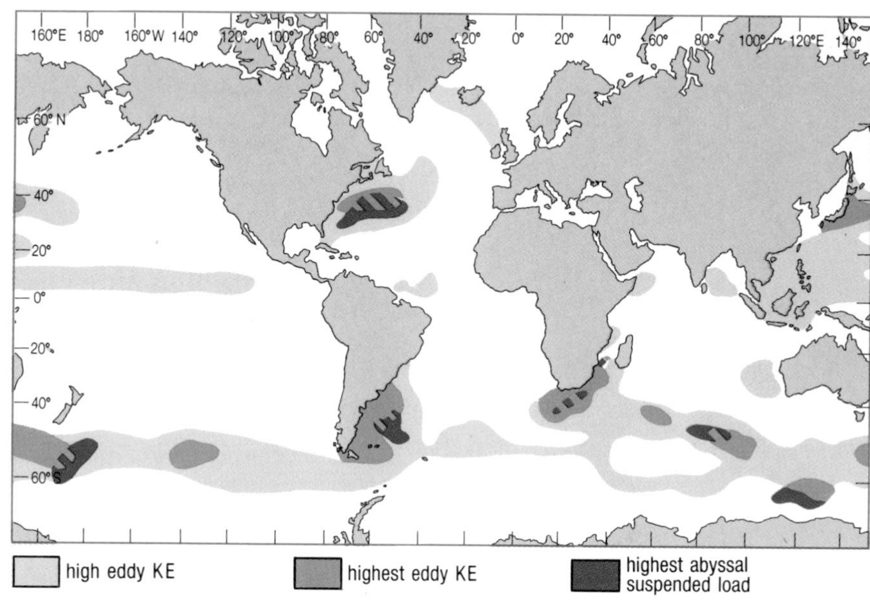

Figure 5.5 Photographs of an abyssal storm event at 4880m depth off Nova Scotia Rise in the north-west Atlantic.

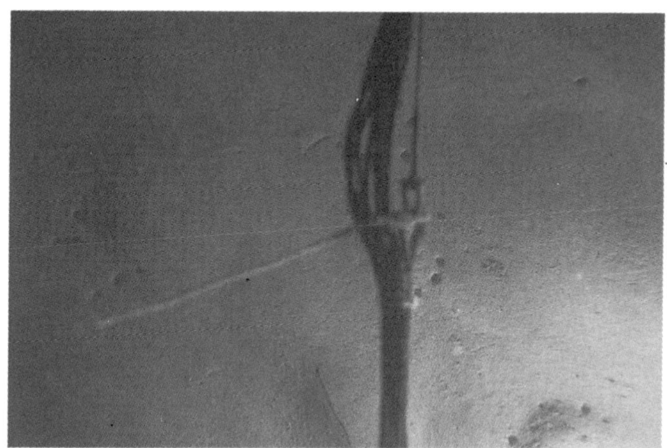

(a) September 30, 1985. The bottom is clean and partially roughened by animal burrows and some tracks. The ridge-shaped ripple feature seen near the bottom centre (just left of the cable) indicates recent bottom activity. The loose cable strand (or rope) shows the current is flowing from right to left.

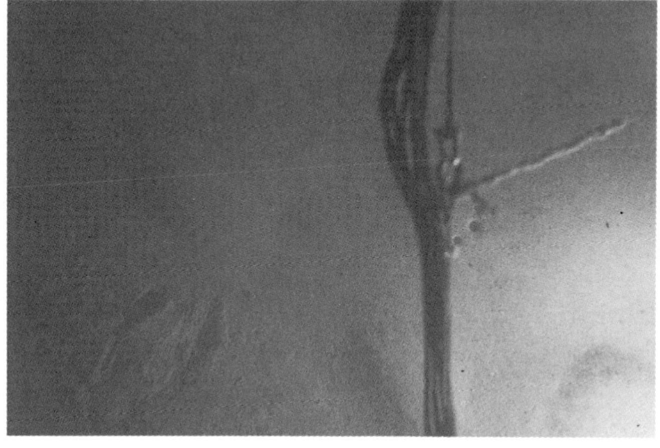

(b) October 31, 1985. An intense abyssal storm has smoothed most of the bed, filling in nearly all the burrows and erasing the tracks and other features. The rough area in the lower left is an impaction crater formed by a piece of falling debris. The current is now flowing from left to right.

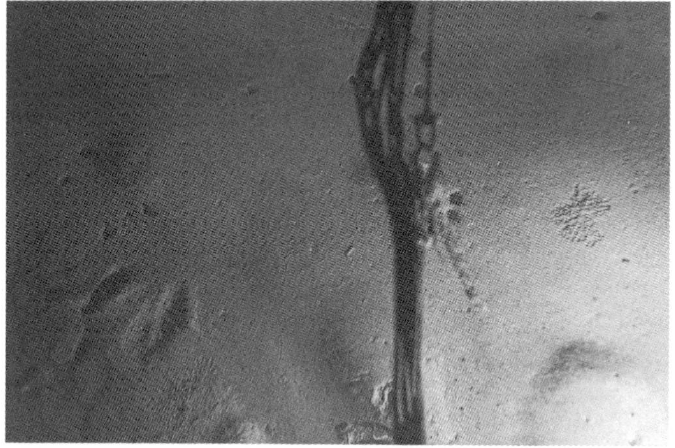

(c) December 27, 1985. Several weeks later, the burrows have been re-opened and there is a general roughening by animal tracks. There has also been some current activity, with 'clouds' of sediment being advected past—some of which has been deposited, smoothing the edges of the impaction crater. The current is now flowing from top to bottom of the picture.

The erosive power of abyssal storms should not be underestimated. For example, at a location beneath the Gulf Stream it has been observed that in less than a year several centimetres of mud and silt have been eroded, and a new deposit (also several centimetres thick) laid down in its place. Sediment lifted up in this way may well reinforce that already being carried in suspension as the yet undeposited contribution from turbidity currents. Concentrations of sediment suspended by abyssal systems can be as much as $5\,g\,l^{-1}$, compared with the normal load in the nepheloid layers of 0.2–$0.5\,g\,l^{-1}$. Abyssal storms may even have implications for the routeing of deep-sea cables: slides, slumps, debris flows and turbidity currents (Chapter 4) are normally seen as their principal enemies, but it seems possible that some stormy areas of the oceanic abyss could also prove hazardous.

As a consequence of turbulent motions near the deep ocean floor, there is a well-mixed *benthic boundary layer*, some tens of metres thick (analogous to what occurs in shelf seas). The layer affects biological, chemical and geological processes at and near the ocean floor and influences the way these processes 'communicate' with the ocean interior. For example, benthic fronts develop in the layer; there can be enhanced mixing downstream of seamounts; and this is the most turbid part of the nepheloid layer, with increased amounts of (re-suspended) organic matter available to the benthos.

5.2 AUTHIGENESIS AND DIAGENESIS IN DEEP-SEA SEDIMENTS

Chemical reactions that form new mineral phases on the sea-bed, either by direct precipitation from seawater or by the alteration of pre-existing material, come under the general heading of **authigenesis**. The term literally means 'self-originating' and it was originally applied at a time when it was thought that the chemical reactions were entirely inorganic, that is, they occurred without the intervention of marine organisms. This is almost certainly not the case with all such reactions, but the name has stuck, and it is important to emphasize that the term is now applied to reactions that occur essentially at the interface between seawater and the sea-bed. However, it also encompasses the inorganic precipitation of mineral phases in the water column, for example, the formation of non-biogenic calcium carbonate in certain areas (e.g. the warm shallow waters of the Bahamas Banks), and the formation of particulate manganese oxide in the dispersed plumes from hydrothermal vents (*cf.* Section 2.2.3).

Much of the montmorillonite in pelagic clays is authigenic rather than terrigenous. That is because seawater rapidly hydrates, devitrifies and decomposes the glassy volcanic material in the volcanogenic component of marine sediments and in the outer rind of lava pillows (Figure 2.2(b)). Montmorillonite clay is a product of such reactions. The rest of the montmorillonite in pelagic clays is land-derived (Section 1.1.2) and this terrigenous component is provided by atmospheric weathering of rocks of basaltic composition on land. There is less of the terrigenous component in the Pacific, where volcanism is widespread and land-derived sediment is trapped in trenches.

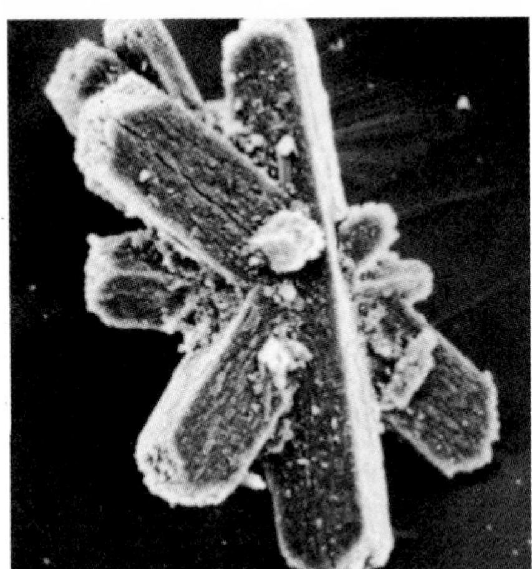

Figure 5.6 Electron micrograph of a cluster of phillipsite crystals recovered from Pacific deep-sea sediments. Scale bar 25μm. The formula of phillipsite is $(K,Na,Ca)_3(Al_3Si_5O_{16}).6H_2O$.

It is unlikely that clay minerals of the other three main groups (kaolinite, chlorite, illite, Section 1.1.2) are either produced or significantly altered by authigenic reactions on the sea-bed. Their distribution in pelagic sediments is consistent with what is known of the latitude belts in which the clays originate (kaolinite in low latitudes, chlorite in high latitudes). This would not be the case if these clays were the products of authigenic reactions. Moreover, radiometric dating of clay minerals in pelagic sediments gives relatively old ages, appropriate to the continental source regions from which the clays probably derived. New phases formed by authigenesis would be younger.

Another authigenic product of reactions between seawater and basaltic glass is *phillipsite*, a potassium-bearing member of a group of hydrated aluminosilicate minerals called *zeolites* (Figure 5.6). Authigenic minerals such as montmorillonite and phillipsite are formed by the alteration of a pre-existing solid (basaltic glass); thus, they contrast with newly produced authigenic phases such as manganese and iron oxides, which are the subject of the next Section.

5.2.1 MANGANESE NODULES

Manganese nodules are probably the most widely known of the authigenic deposits found on the deep ocean floor, even though they are among the most enigmatic. They have rounded shapes and range in size from micronodules only a few μm in size up to nodules more than 10cm across and a kilogramme or more in weight. Most are about the size of a tennis ball or a little less. Concentrations of nodules on the sea-bed can reach $25\,kg\,m^{-2}$.

Manganese nodules were first recovered from the Atlantic Ocean floor, close to the Canary Islands, during the *Challenger* expedition (Figure 5.7). Since then, they have been mapped in all oceans, except the Arctic, mainly by photographing the ocean bed. Sometimes their occurrence is patchy, but elsewhere they may form rich nodule fields (Figure 5.8), and in some areas slabs or 'pavements' extend over several square metres and

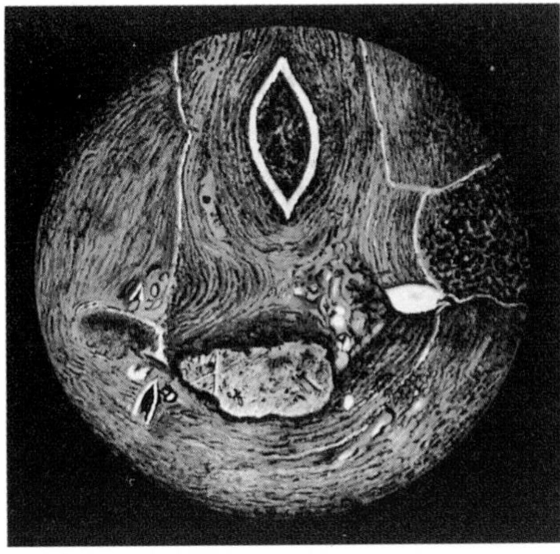

Figure 5.7 Drawing of a section cut through a small manganese nodule, about 2cm across. The nodule has grown round a shark's tooth (top centre) and a volcanic fragment (bottom centre), and has incorporated other particles as it grew. This was recovered by the *Challenger* expedition from a depth of 2375 fathoms (*c.* 4350m) in the South Pacific.

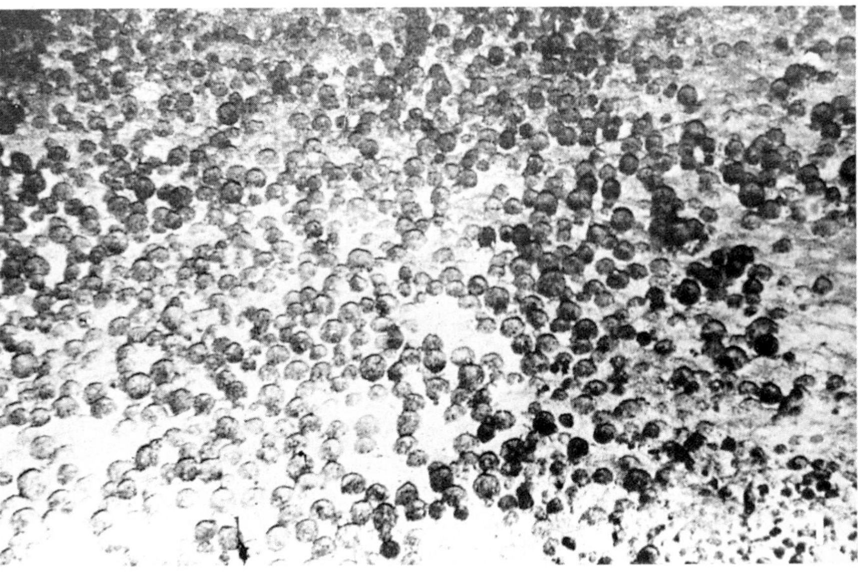

Figure 5.8 Typical appearance of a rich manganese nodule field in the deep ocean.

weigh several tonnes. In addition, there are manganese-rich coatings on the surfaces of igneous (usually volcanic) rocks exposed on the ocean floor. The distribution of nodules in the Pacific and Atlantic Oceans is shown in Figure 5.9.

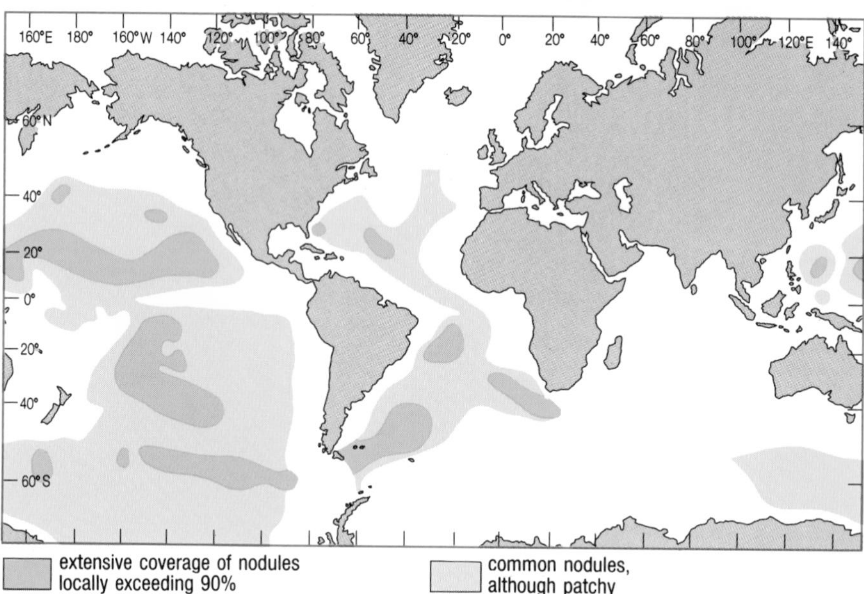

Figure 5.9 A generalized map of the distribution of manganese nodules in the Atlantic and Pacific Oceans.

QUESTION 5.2 Compare Figure 5.9 with Figure 2.8. Is there any general correlation you could make between those regions where nodules are most abundant, and conditions on the deep sea-floor? Would you expect to find nodules in the Indian Ocean?

Manganese nodules are most abundant where sediment accumulation rates are lowest, usually less than $5\,\mathrm{m\,Ma^{-1}}$. Many of the clays in which they are found are the so-called red clays (Section 1.1.2) which indicate well-oxygenated bottom waters. Nodule abundances decrease towards the ocean basin margins where sedimentation rates increase because of high inputs of terrigenous sediment. Nodules are also less common directly beneath the equatorial belt of high productivity in the Pacific, where the input of biogenic material is high (Figure 1.4).

The nodules are mainly mixtures of manganese and iron oxides and hydroxides which have been deposited in concentric layers around a nucleus of some sort. This may be of biological origin (e.g. a shark's tooth, Figure 5.7), but is more commonly a volcanogenic fragment (Figure 5.10). Rates of growth of deep-sea manganese nodules are estimated to be in the order of a few millimetres per million years, which is three orders of magnitude less than sedimentation rates (a few metres per million years) of the deep-sea sediments among which they are found, a topic to which we return shortly.

The composition of manganese nodules

Table 5.1 shows that there are appreciable differences in the average compositions of nodules from the three main ocean basins. The concentrations of nickel, cobalt and copper are of particular interest, because their combined concentrations can exceed 3%, which makes them potential ores for these metals. Much of the impetus for research

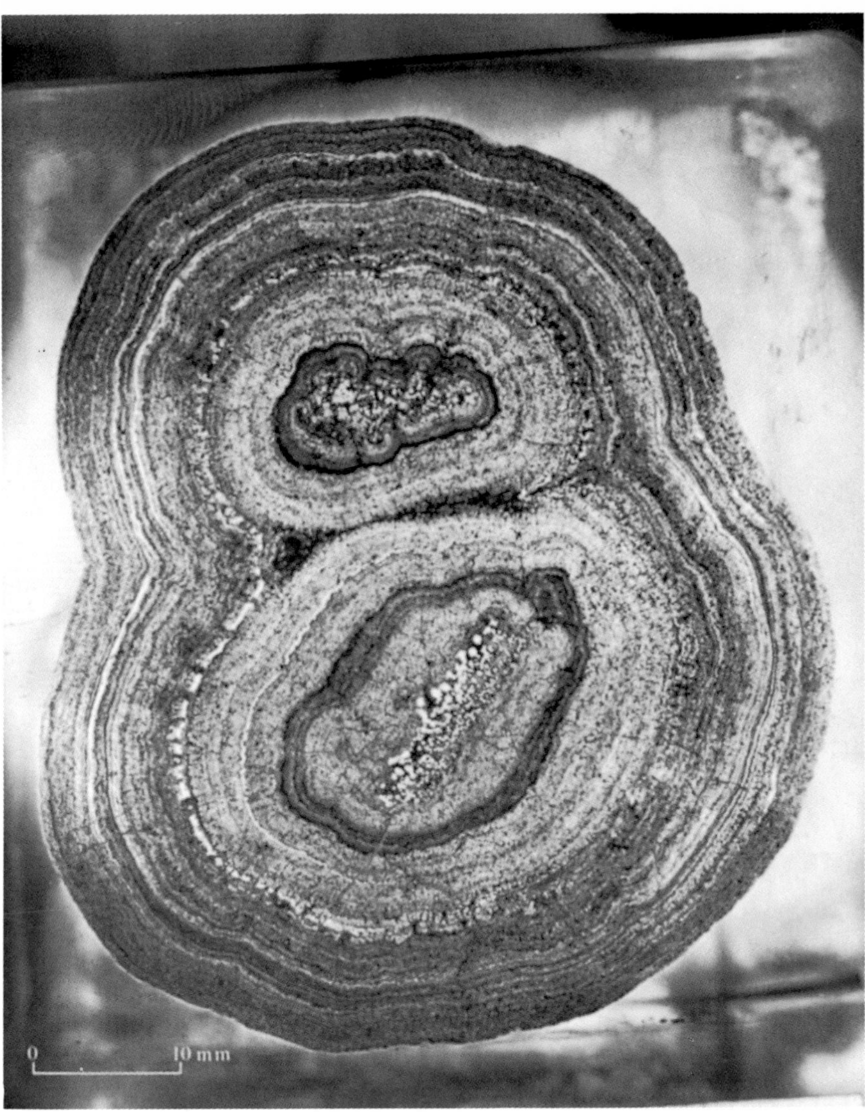

Figure 5.10 Cross-section through a manganese nodule, showing concentric growth about two nuclei which are probably fragments of altered volcanic rock (basalt).

into manganese nodules has been fuelled by commercial interests eager to exploit this resource of the deep sea-bed.

Table 5.1 The average abundances of manganese (Mn), iron (Fe), nickel (Ni), cobalt (Co) and copper (Cu) in manganese nodules from the Atlantic, Pacific and Indian Oceans, and the average for all three oceans. The numbers represent the weight % of each metal.

Element	Atlantic	Pacific	Indian	Average for all three oceans
Mn	16.18	19.75	18.03	17.99
Fe	21.2	14.29	16.25	17.25
Ni	0.297	0.722	0.510	0.509
Co	0.309	0.381	0.279	0.323
Cu	0.109	0.366	0.223	0.233

QUESTION 5.3 On the basis of Table 5.1 alone, in which ocean basin would you concentrate your search for the 'richest' nodules? Would you necessarily be right to do so?

The conditions which determine variations in nodule composition change rather gradually throughout the oceans. While nodules from within a particular area will all have roughly the same composition, there may be considerable differences in the average compositions of nodules from different regions. The extent to which metals are taken up may be determined in part by the crystalline structure of the manganese oxides and hydroxides which are the major component of the nodules.

The growth of manganese nodules

This topic presents a number of interesting problems. For example, how do the nodules grow and how do they incorporate the various metals into their structure? Is the growth process wholly inorganic, or are biological agencies involved? Why do nodules not become buried, if they grow so slowly in comparison with the accumulation rates of deep-sea sediments?

Figure 5.11 shows some results of an analytical scan across a small manganese nodule, and gives you an idea of the variability of the internal composition of nodules. Individual details are not important, but two points are noteworthy. First, there is considerable variation associated with the concentric layering, and some correlation between the patterns of variation for the three elements shown; and secondly, the non-authigenic (pre-existing) nucleus or core displays no such variability, though it has evidently been somewhat enriched in Mn and Cu (but not in Ni).

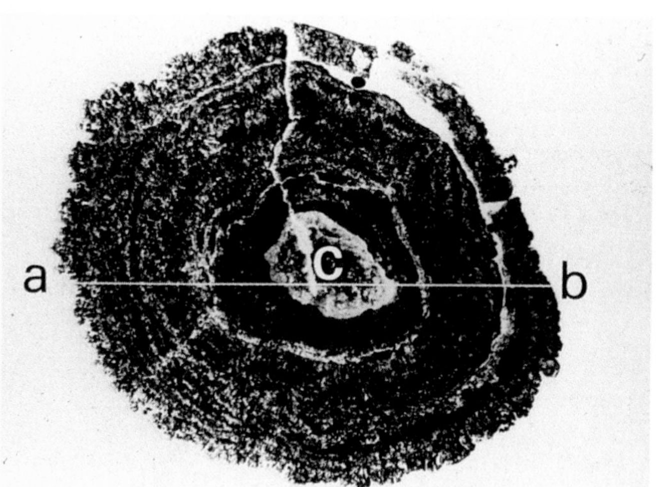

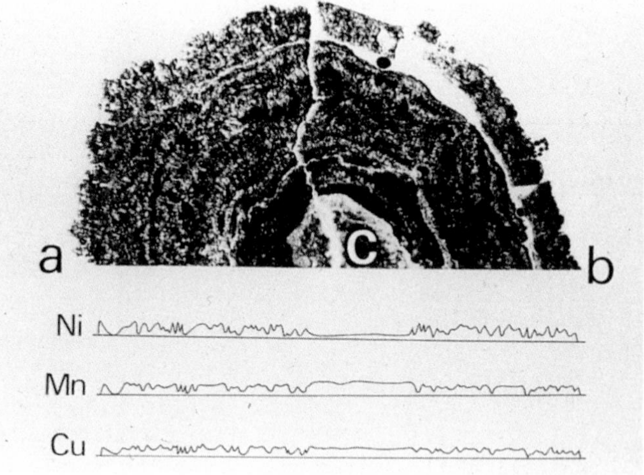

Figure 5.11 Cross-section through a manganese nodule (left), analysed by a continuous scan (right) along the line a–b for nickei (Ni), manganese (Mn) and copper (Cu). The arbitrary vertical scales give relative amounts of each metal along the scan, and they are *not* the same for each metal. The core (c) is free of compositional oscillations and is presumably the original nucleus, probably altered (weathered) volcanic rock. This is a small nodule, about 2cm across.

Rates of growth of manganese nodules and crusts have been measured with the help of natural radio-isotopes (e.g. ^{230}Th, ^{10}Be). Accretion rates can vary from about 1 to 6mm per million years; and changes in growth rate and in the composition and detailed structure of successive layers have been correlated with past climatic events, in particular the glacial fluctuations that commenced during late Tertiary times (the last 15Ma or so). The highest growth rates were recorded for the period from about 125000 years ago to the present; but there seems to be no hard and fast correlation between growth rate and climatic conditions.

It is fairly obvious from Figure 5.8 that nodules mostly lie partly buried in the sediment, rather than completely on top of it. If they are to grow with a more or less sub-spherical concentric shape, it would seem to follow that accretion must occur both above and below the sediment–seawater interface; that is, both from the overlying seawater and from pore waters between the sediment particles on the sea-bed.

Does Figure 5.12 suggest that the rate of accretion on the upper surface of a nodule might be different from that on the lower surface?

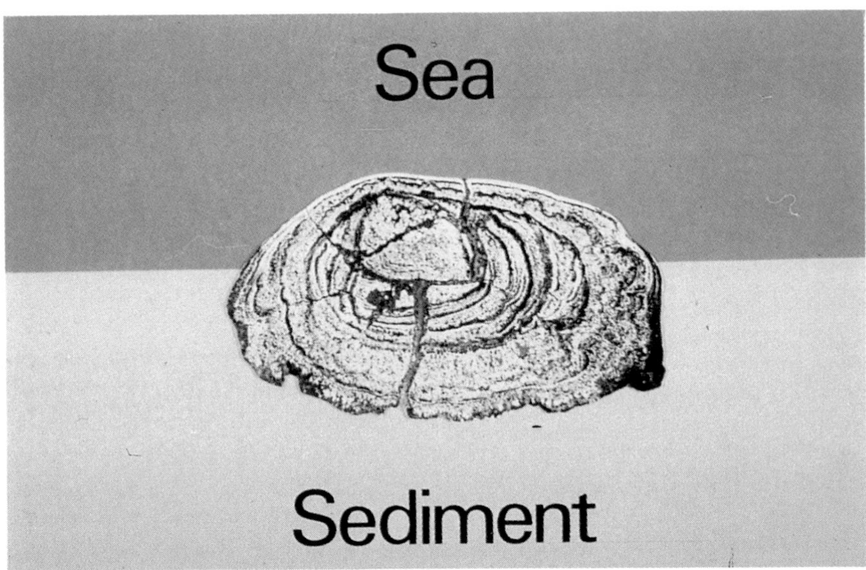

Figure 5.12 A nodule cross-section drawn in 'growth' position at the sea-bed. For discussion see the text.

On this nodule, the lower layers are considerably thicker than those above the sediment–seawater interface, so accretion rates were greater below the interface. Moreover, analysis of this particular nodule showed that both the Fe:Mn ratio and the concentrations of Cu and Ni were greater in the bottom layers than in the top layers. The importance of fluxes of elements from pore waters in the surrounding sediments has been widely argued in hypotheses of nodule formation, and compositional differences between the upper and lower parts of nodules are relevant to such arguments.

Closer inspection of nodule surfaces reveals them to be rich mini-ecosystems in their own right (Figure 5.13). They support diverse populations of mainly protozoan animals, including Foraminifera, many of which use organic 'glues' to stick fine-grained detrital particles together, forming a variety of skeletal structures, some of which can also be found preserved within the nodules. These agglutinating or 'arenaceous' animals are found mainly on nodule surfaces above the sediment–seawater interface; below the interface, the animals are softer and more gelatinous. While these organisms may themselves not contribute directly to the growth of nodules, it is possible that the basic framework consists of biogenic tubes and other structures initially built round the original nucleus, and that this framework then becomes filled in by precipitation of manganese and other metals to form the actual nodules.

The shape and structure of most nodules suggests that they must grow more or less uniformly from both top and bottom. Although there are reports of nodules being recovered in drill cores, it is probably fair to say

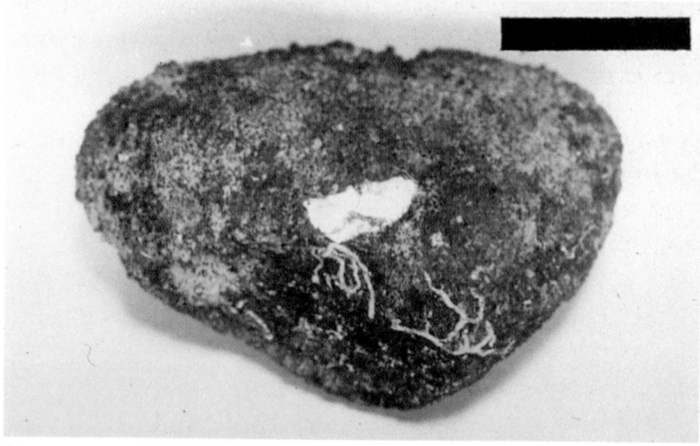

(a)

(b)

Figure 5.13 (a) The surface of a nodule showing a sponge colony (white patch in centre) and filamentous benthic Foraminifera (white 'ribbons', lower centre), as well as numerous smaller organisms in lighter patches that cannot be resolved at this scale.

(b) Electron micrograph showing interconnected 'domes' and tubes formed by arenaceous protozoans. The round spiny object in the lower centre is a planktonic radiolarian test that has sunk from the surface.

that the great majority lie *at* the sediment–seawater interface (Figure 5.8). Nodule fields can thus be very extensive, but they must in general be very thin—little more than the diameter of the average nodule. Relatively few nodules appear to be buried, yet the sediments are accumulating very much faster—*metres* per million years—than the nodules themselves are growing—*millimetres* per million years.

What does this tell you about the ages of the nodules compared with the surrounding sediment?

The nodules must be considerably *older*. Several possibilities have been suggested to explain this riddle, but the evidence for them is equivocal. One suggestion is that they are periodically rolled along the sea-floor by bottom currents which also erode the surrounding sediments. The high concentrations of nodules in the Southern Ocean (Figure 5.9) might be attributed to the incidence of abyssal storms beneath the **Antarctic Circumpolar Current** (*cf.* Figure 5.4). However, there is no evidence of significant bottom current activity in other parts of the deep oceans where nodules are just as plentiful, if not more so.

Figure 5.14 Photograph by Bruce Heezen of large manganese nodules half buried in sediment and showing evidence of biological activity: there is a coil of holothurian faeces in the centre left of the picture.

A second possibility, perhaps more appealing to the imagination, is that the nodules are constantly being nudged by foraging benthic organisms, which roll the nodules over and thus keep them at the sediment surface. The low concentrations of organic matter in the pelagic clay accumulations of nodule fields (*cf*. Figure 2.8) means that the rich fauna on the nodule surfaces (Figure 5.13) could be 'grazed' by larger benthic animals. A correlation between the occurrence of nodules and the presence of benthic animals has been observed in some ocean floor photographs (Figure 5.14). However, there are very few large animals at abyssal depths, and in any case such an explanation cannot account for manganese oxide pavements.

Manganese nodules in other environments

Long before the time of the *Challenger* expedition, manganese nodules were known to occur in freshwater lakes. Indeed, they were mined as 'bog ores' in Scandinavia during medieval times.

Table 5.2 shows that marine manganese nodules are not found only at abyssal depths. The nodules that grow on seamounts and ocean ridges are notable for their higher concentrations of cobalt relative to nickel and (especially) copper, but their manganese and iron contents and Mn:Fe ratios are not very different from those of abyssal nodules (the average composition of which is the same as in the last column of Table 5.1).

QUESTION 5.4 From Table 5.2, in what major respects do manganese nodules from continental margin environments differ from the rest?

Table 5.2 Average composition of manganese nodules from different environments (weight %).

Element	Seamounts	Active ridges	Continental margins	Abyssal depths
Mn	14.62	15.51	38.69	17.99
Fe	15.81	19.15	1.34	17.25
Ni	0.351	0.306	0.121	0.509
Co	1.15	0.400	0.011	0.323
Cu	0.058	0.081	0.082	0.233
Mn/Fe	0.92	0.81	28.9	1.04

Like their abyssal equivalents, continental margin deposits of manganese and iron oxides and hydroxides can take on a variety of forms, from spheroidal nodules through discs to extensive slabs, and are found either on the sea-floor in contact with the overlying seawater, or in the uppermost layers of sediment. Unlike abyssal nodules, however, they form relatively rapidly: growth rates are of the order of 0.01–1 mm per year, compared with around 1–5 mm per million years for deep-sea nodules. They are also much less rich in the metals that make deep-sea nodules so interesting as potential ore deposits.

Not only do continental margin sediments accumulate much more rapidly than deep-sea sediments, they also contain more organic matter. Because the decay of buried organic matter quickly uses up the available oxygen, anoxic conditions develop within the sediments, beneath a thin surface layer of oxidized sediments. The interface between the anoxic zone and the oxidized surface layer migrates upwards as more sediment is deposited. As outlined in Section 2.5.3, manganese is more soluble in anoxic than in oxidizing environments. Therefore, manganese-bearing phases precipitated under oxidizing conditions on the sea-floor are redissolved after burial in the anoxic environment of the sediment pile. The dissolved manganese migrates back to the sediment surface in pore waters and is re-precipitated there (of course, there is also direct precipitation of manganese from the overlying seawater). The relatively rapid formation of these nodules is thus primarily a function of high rates of sedimentation: the interface between anoxic and oxidized sediments is much closer to the sea-bed than in pelagic environments, where sedimentation rates are low and there is less organic matter—as a result, dissolved manganese cannot be supplied from pore waters in the way outlined above.

The comparative lack of metals such as copper, cobalt and nickel in continental margin nodules is another feature that can be related to the sedimentary environment. The nodules of continental margins grow very much faster than those in the deep oceans—by several orders of magnitude, as we have seen. However, the rate of precipitation of the other metals is probably everywhere about the same, and so they are greatly 'diluted' in the faster-growing nodules.

For completeness, we should note that not all continental margin nodules are rich in Mn relative to other metals. For example, on the Blake Plateau off Florida, nodules have Mn:Fe ratios of around unity and show some enrichment of copper, cobalt and nickel.

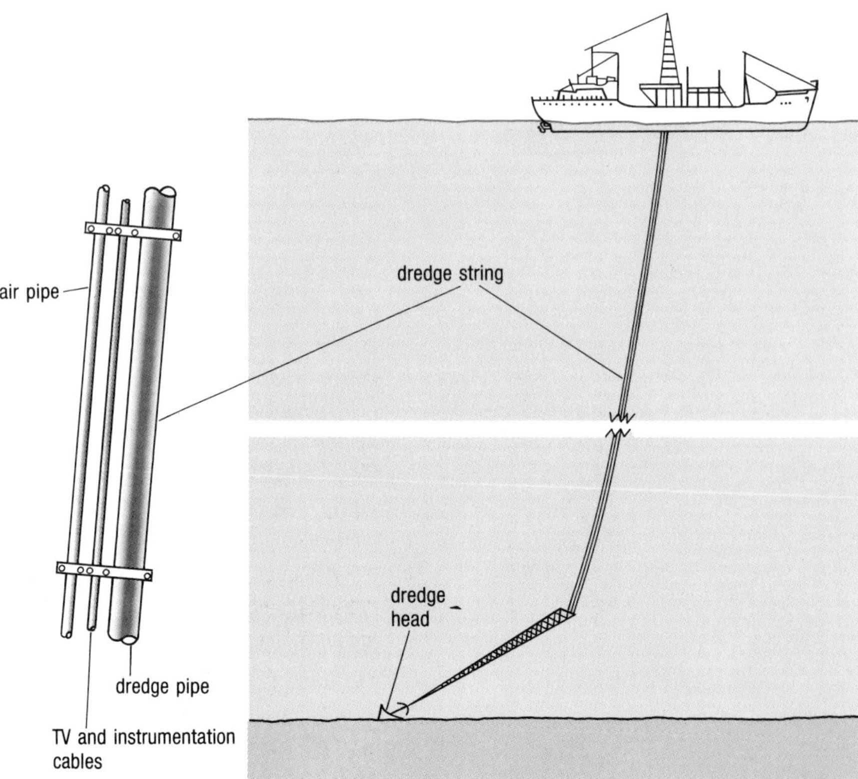

Figure 5.15 The suction method for recovering manganese nodules from the ocean floor.

The exploitation of deep-sea nodules

As already noted, it is the combined concentrations of copper, nickel and cobalt (Table 5.1) that make nodules such an attractive proposition for mining companies. There are three major obstacles to exploitation:

1 The technological problems of extracting a thin layer of nodules from the sea-bed at depths of about 5 km are formidable. One possible method, which involves simply sucking the nodules off the bed, is illustrated in Figure 5.15.

2 Major environmental disruption could result not only from the dredging or 'hoovering' of large tracts of sea-floor, but also from the disposal of the large volumes of muddy 'waste' sediment sucked up along with the nodules.

3 Who owns the high seas? The short answer is nobody, and there is no provision in the Law of the Sea for establishing mineral rights to the sea-bed beneath the open oceans. One of the principal intended functions of the International Sea-Bed Authority is to regulate deep-sea mining operations so that no single nation, consortium or operator obtains an unfair economic advantage. However, prices of these metals on world markets are unlikely to rise sufficiently to make deep-sea nodule mining an economic proposition for the immediate future, so there may be time for the legal situation to be resolved.

5.3 DIAGENESIS

Diagenesis is the term applied to the chemical changes that occur within sediments, through interaction with pore waters, as they become compacted after burial and eventually lithified and recrystallized to form sedimentary rocks. Diagenetic changes can also affect volcanic rocks when they become buried by sediments.

It is difficult to draw hard and fast lines between authigenic and diagenetic reactions. Here, we define authigenic reactions as those occurring at the sediment–seawater interface (Section 5.2), and diagenetic reactions as those occurring beneath the interface. However, authigenic processes initiated at the sediment–seawater interface commonly continue after burial, as reactions between sediments and pore waters—when they should strictly speaking be termed diagenetic, even though they may still involve the formation or alteration of clay minerals and zeolites. So, we cannot always be sure just when an authigenic mineral phase becomes a diagenetic one, and you may find these terms used in apparently contradictory ways in the literature, simply as a consequence of this uncertainty—in short, it is a matter of semantics.

Once we get more than a few tens of centimetres below the sediment–seawater interface, the rate of exchange between seawater and interstitial pore waters in the sediments decreases considerably. Reactions between sedimentary particles and trapped pore waters can be reliably defined as diagenetic.

One of the better known and more obvious diagenetic changes is the gradual long-term lithification of calcium carbonate sediments to form chalk or limestone. The recrystallization commonly involves some replacement of calcium by magnesium to form the mineral *dolomite*,

$(CaMg)CO_3$: magnesium is removed from solution in the pore waters, while calcium goes into solution.

Siliceous biogenic sediments are also gradually lithified to form a very hard rock called *chert*, but otherwise undergo negligible change during diagenesis, except for the loss of some silica to pore waters. (Varying amounts of silica as detrital quartz occur in almost all sediments, but this also undergoes very little change during diagenesis, beyond some solution and recrystallization along grain boundaries.)

For pelagic clays, chemical changes are very slow, so long as the environment remains oxidizing, as it generally does in at least the top few tens of centimetres in most sediments. This is not really surprising, because the oxygenated environment at the sea-floor is not very different from that in which continental weathering produces the detrital minerals in the first place. Temperatures are generally lower and pressures a good deal higher, but the chemical environment is still oxidizing, even though the minerals are in contact with seawater instead of the atmosphere.

However, at greater depths within the sediment, where anoxic conditions do develop, the chemical environment rapidly becomes quite different from that of continental weathering. This is marked by a horizon within the sediments at which new diagenetic reactions between pore waters and clay minerals begin to take place to a significant extent. As we saw in the previous Section, anoxic conditions are likely to be found a short distance below the sea-bed in sediments near continental margins, where pelagic clays merge into clays that contain material deposited from lutite flows and turbidites. Clays of the continental rise contain much more terrigenous material and organic matter, and accumulate much more rapidly than true pelagic clays; and they tend to 'seal' the surface because of their fine grain size and tendency to be cohesive. It is in these sediments that the most marked changes occur during diagenesis.

An early sign of reducing conditions in the sediment column is the depletion of sulphate in the pore waters (Section 2.5.3). A common diagenetic mineral of anoxic environments is *pyrite* (FeS_2), formed by the reaction of dissolved sulphide (produced by the reduction of sulphate, reaction 2.5) with reduced iron (Fe^{2+}).

Where conditions remain oxidizing within the sediments, the effects of diagenetic reactions become significant only where the sediment pile has become fairly thick and plenty of time has elapsed. We shall consider one example to illustrate the kinds of changes that occur in these circumstances (Figure 5.16).

The top 400m of the sediment sequence collected from DSDP Hole No. 149 in the Venezuela Basin of the Caribbean consists of a variety of sediments that go back to the Middle Eocene (about 50Ma ago), giving a mean deposition rate of 8 metres per million years, which is typical for pelagic sediments. Above the chert band at the base of the sequence (Figure 5.16), the sediments are dominated by biogenic carbonates, but below about 250m there is a significant siliceous (radiolarian) component. As far as non-biogenic components are concerned, material of volcanic origin is common below about 100m, where authigenic montmorillonite is the dominant clay mineral; above 100m, illite and kaolinite are the main clay minerals. There is evidence of some oxygen depletion in the pore waters, but the environment remains oxidizing throughout the sequence.

QUESTION 5.5 (a) Examine Figure 5.16 and determine which of the four constituents represented there become enriched, and which become depleted in the pore waters of these sediments, with increasing depth.

(b) From the description of the sediment sequence given above, can you account for the change in the dissolved silica content of the pore waters with depth?

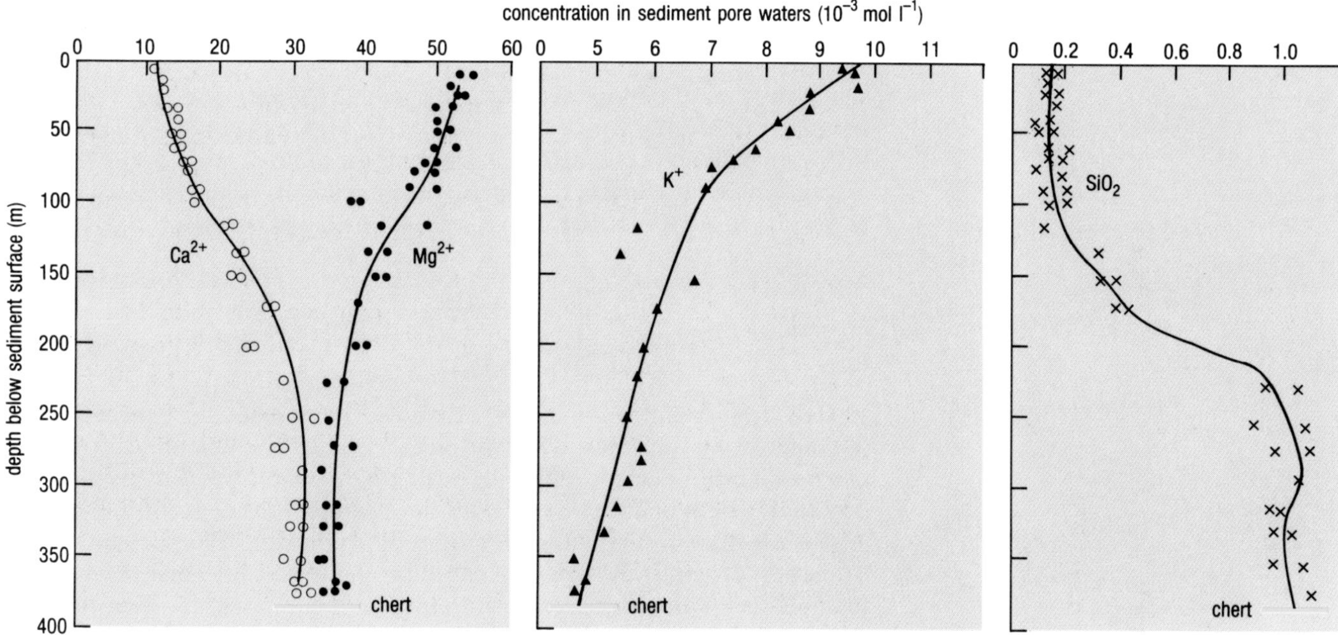

Figure 5.16 Vertical concentration profiles for four constituents in the pore waters of DSDP Hole 149. Note the layer of diagenetic chert at the base of the sediment sequence.

You may also have noticed another interesting feature in Figure 5.16(a): one curve is almost a mirror image of the other, so that the molar sum of Ca^{2+} and Mg^{2+} ions is almost constant in the pore waters throughout the sediment column—it appears that one Ca^{2+} ion is released to solution for every Mg^{2+} ion removed. No evidence of dolomite formation has been found in these sediments, and the explanation advanced for this one-for-one relationship of Ca^{2+} and Mg^{2+} is that it is due to the breakdown of volcanic debris: initial formation of authigenic magnesian montmorillonite continued as a diagenetic process long after the volcanic debris from which it derived was buried in the sediment pile.

What can you say about the behaviour of potassium in this sequence?

The gradient of depletion of potassium in pore waters is greatest in the uppermost 100m, where, as outlined above, detrital illite and kaolinite are the dominant clay minerals. There is evidence that these detrital clays are incorporating potassium (see Section 5.3.1), and their greater abundance near the top of the sequence would thus account for the greater potassium depletion there. That does not preclude the incorporation of potassium by volcanic materials, however, and the presence of zeolite minerals in at least one of the core samples below 100m is evidence that this may be happening (see also Section 5.3.2).

5.3.1 AUTHIGENESIS, DIAGENESIS AND THE MAJOR CONSTITUENTS OF SEAWATER

As you know, several major constituents of seawater have element to salinity ratios that are virtually constant throughout the oceans (Section 2.1); they have long residence times, of the order of millions to tens of millions of years; and they are bio-unlimited elements (Section 2.2.4). The sinks that act to maintain the concentrations of these elements in seawater are therefore likely to be inorganic ones.

Could authigenic or diagenetic reactions provide a sink for any of the major constituents?

The formation of phillipsite and the diagenetic changes affecting clay minerals represent possible sinks for *potassium*. Diagenetic reactions involving both clays and calcareous sediments could provide a sink for *magnesium*. One possible diagenetic reaction involving the removal of potassium from pore waters by reaction with clay minerals is:

$$\underset{\text{kaolinite}}{5\,Al_2Si_2O_5(OH)_4} + \underset{\text{from solution}}{2K^+ + 2HCO_3^- + 4\,SiO_2} \rightarrow \underset{\text{illite}}{2KAl_5Si_7O_{20}(OH)_4} + 7H_2O + 2CO_2 \quad (5.1)$$

At first sight, it may seem unlikely that diagenetic reactions between sediment and pore waters deep within the sediments could affect the composition of seawater. However, when pore waters become relatively depleted in potassium or magnesium, for example, the concentration gradient formed will result in these ions diffusing along the gradient from the region of relatively high concentration (seawater) into that of relatively low concentration (pore water). In Figure 5.16, for example, both potassium and magnesium are being continually removed from seawater into pore waters as a result of the diagenesis. Clearly, this process of removing dissolved constituents from seawater is extremely slow, but then the mean oceanic residence times of elements such as magnesium and potassium are correspondingly very long (Section 2.2.4). Conversely, Figure 5.16 suggests that silica and calcium, which are released into solution in the pore waters during diagenesis, diffuse back up to the sediment–seawater interface to rejoin the oceanic cycles.

5.3.2 DIAGENESIS AND ZEOLITES

Phillipsite (Figure 5.6) is an authigenic mineral, and it tends to decrease in abundance with increased depth in the sediment pile as a result of alteration and decomposition. Diagenetic reactions can form other zeolites, e.g:

Clinoptilolite (Figure 5.17(a)) is rare in surface sediments, but becomes more abundant with depth in the sequence, especially where the sediments contain calcareous and siliceous components. The Si:Al ratio in clinoptilolite is about 4, compared with about 2 in phillipsite, and it contains more calcium and less of the alkali elements (Na and especially K) than phillipsite.

Analcime (Figure 5.17 (b)) is another zeolite that is rare in surface sediments and is relatively common in older sediments, where it appears to have formed from the diagenetic alteration of other zeolites. Its structure accommodates quite a lot of sodium, and it is less hydrated than

(a)

(b)

Figure 5.17 Electron micrographs of crystals of: (a) clinoptilolite (scale bar 2 μm). The formula of clinoptilolite is (Ca,Na) $(Al_2Si_7O_{13}).6H_2O$. (b) Analcime (scale bar 10 μm). The formula of analcime is $Na(AlSi_2O_6).H_2O$.

most other zeolites; this is consistent with its formation fairly deep down in sediment sequences, where compaction expels water from the pore spaces between sediment particles. The removal of sodium from pore waters to analcime thus provides a long-term slow-acting sink for the sodium in seawater.

5.3.3 THE OXIC–ANOXIC INTERFACE

In the example discussed in Figure 5.16, the pore waters are described as being oxygenated throughout the sequence, albeit with some depletion of oxygen. In regions where sediments accumulate rapidly and there is a significant amount of organic carbon present, an oxic–anoxic interface can develop relatively close to the surface.

Research into diagenesis in turbidite sequences on continental rises and adjacent abyssal plains was intensified during the early to mid-1980s. This was because of a proposal that nuclear and other hazardous wastes might be disposed of by burial deep within these thick sediment sequences. A major part of the research was to investigate the extent of element mobility within the sediments, via the pore waters. Clearly, it was important to know whether toxic substances might escape into the water column, and, if so, at what rates.

Huge amounts of data were collected, and we have space to look only at one tiny aspect: the uppermost parts of a couple of sediment cores collected from the abyssal plain area off north-west Africa. The sediments contain about 63% of calcium carbonate (mainly coccolith calcite), the rest being mainly clay. They are believed to have accumulated originally on the upper continental slope, where they became unstable and were transported by a turbidity current to the deep sea-floor, which appears to be close to the CCD in this area (Figure 3.4). The cores have a pale brown top layer overlying pale grey sediment with a sharp colour transition at about 20cm depth below the sea-bed.

Composition–depth profiles for major elements in the sediment itself show that generally there is very little change of composition across the colour transition (Figure 5.18(a)). However, when we look at the results obtained when the pore waters were analysed for redox-sensitive species, a somewhat different picture emerges (Figure 5.18(b)).

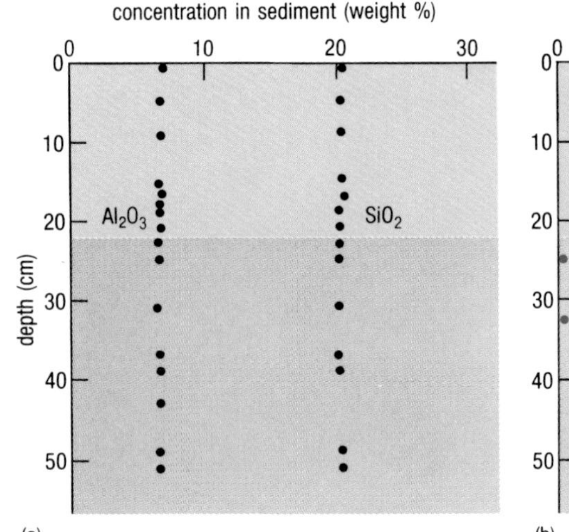

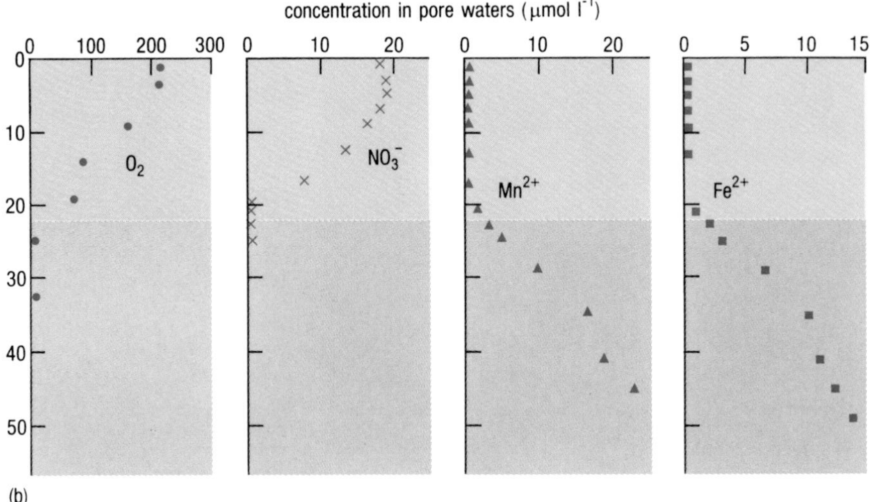

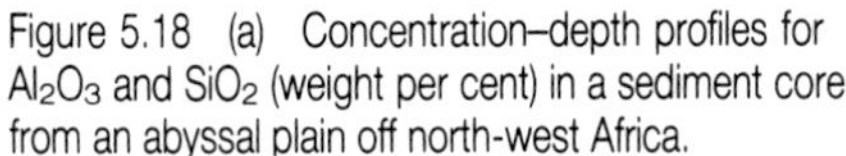

Figure 5.18 (a) Concentration–depth profiles for Al_2O_3 and SiO_2 (weight per cent) in a sediment core from an abyssal plain off north-west Africa.

(b) Concentration–depth profiles for dissolved constituents in pore waters in a sediment core from the same sediment sequence as in (a).

QUESTION 5.6 (a) Why does Figure 5.18(a) enable us to be fairly confident that there is little significant change in sediment composition with depth in the sequence sampled?

(b) Why does the oxygen profile in Figure 5.18(b) enable us to conclude with confidence that the brown–grey colour transition must be the oxic–anoxic interface in the sediment pile? Why does the nitrate profile have the same form?

(c) Why do both Mn^{2+} and Fe^{2+} increase downwards away from the transition?

The behaviour of uranium is of interest in this context. There are small amounts of this element in nearly all rocks, so there is bound to be some in terrigenous sediments (including clays) derived from rock weathering. Uranium is more soluble in its oxidized (U^{6+}) form than in its reduced (U^{4+}) form. It follows that uranium should be released into solution from sediment particles in the upper oxidized zone and immobilized in insoluble form in the underlying reduced zone.

Does Figure 5.19 bear out these general predictions?

Concentrations of both organic carbon and uranium increase sharply across the colour transition. The high value for organic carbon below the transition testifies to the lack of oxygen there, and the consequent reducing conditions must be responsible for the precipitation of uranium. The peak in the uranium profile just below the transition is interpreted as being due to the oxic–anoxic interface itself migrating *downwards* as an oxidation front; which means that oxygen is diffusing *into* the sediment from the overlying seawater. Analyses of cores penetrating deeper into the sedimentary sequence reveal other uranium peaks, deeper down. The interpretation is that each successive turbidite sequence has sealed off and fossilized similar elemental profiles in the underlying sediment.

Does this suggest that sediment deposition on abyssal plains is intermittent or more or less uniform?

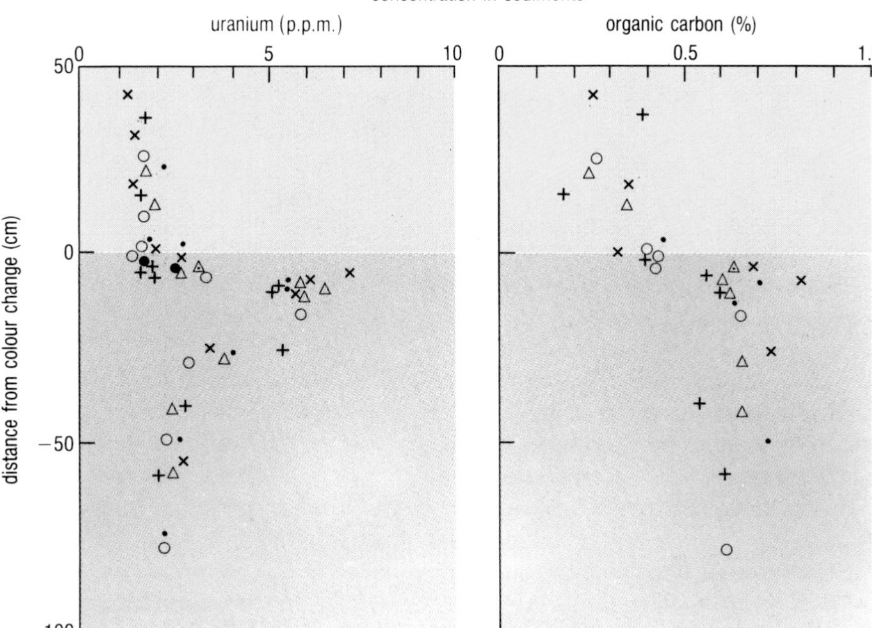

Figure 5.19 Concentration–depth profiles for uranium and organic carbon in several sediment cores from the same sediment sequence as in Figure 5.18. Each symbol corresponds to a different core.

The rain of pelagic sediment descending to the abyssal plain is continuous and more or less uniform, but it is violently interrupted at irregular intervals of perhaps hundreds to thousands of years by turbidity currents which—even in abyssal plain regions—can deposit as much sediment in a matter of days or weeks as the 'normal' pelagic sedimentation supplies in a million years.

We have provided only a small glimpse into one aspect of a research programme that went on for several years. No firm conclusions were reached about whether or not continental rise or abyssal plain regions might be suitable for waste disposal. The indications were favourable, although the research into what was thought to be a monotonous and quiet abyssal plain region has shown that far more goes on there than was previously suspected. Funding of the project ceased in 1986–7, even though the need to find a solution to the disposal of nuclear waste has not gone away.

Hydrocarbon formation

We have seen that in anoxic conditions the decomposition of organic matter can be achieved by using sulphate or nitrate as oxidizing agents (e.g. reaction 2.5). In the anoxic environment of a thick pile of sediments, these sources of oxygen can become used up and direct bacterial oxidation of organic matter can occur:

$$2CH_2O \rightarrow CH_4 + CO_2 \qquad (5.2)$$

and under even more reducing conditions, carbon dioxide itself is decomposed:

$$CO_2 + 4H_2 \rightarrow CH_4 + 2H_2O \qquad (5.3)$$

Both these reactions lead to the formation of methane and other light hydrocarbons (e.g. ethane, ethylene, propane) and occur wherever there are large accumulations of anaerobic organic matter, as in marshes, rubbish dumps and manure heaps. They are simple versions of the type of reaction that is important in the development of petroleum accumulations. During the 1980s, hydrocarbon seeps were found along continental margins, surrounded by communities of animals similar to those colonizing hydrothermal vents along ocean-ridge axes, and nourished by the hydrogen sulphide and hydrocarbons escaping from the sea-bed there.

Methane and other light hydrocarbons have long been known to form gas hydrates (e.g. CH_4nH_2O) under certain conditions of low temperature and elevated pressure. Conditions appropriate to the formation and stability of gas hydrates (also known as clathrates) exist beneath the upper parts of the continental rise, where sediments are thickest and the content of organic matter is likely to be greatest. During the late 1980s, research into these gas hydrates was intensified, as it was realized that they may contain large reserves of hydrocarbon fuels. The technology of petroleum extraction in deep water is improving all the time, and such deposits could have great commercial potential.

5.4 SUMMARY OF CHAPTER 5

1 Deep-sea sediments are disturbed and mixed (bioturbated) by animals moving over and through them in the search for food. Bottom currents can make bed forms such as ripple marks on the surface, and they can resuspend large amounts of sediment. In regions of the ocean where there are strong western boundary currents, abyssal storms lead to erosion of bottom sediments. In the lowermost few hundred metres, the water column is turbid with suspended sediment and is called the nepheloid layer.

2 Authigenesis—the formation of new minerals at the sea-bed—includes the formation of montmorillonite clay, the zeolite phillipsite, and manganese nodules. The latter are mainly spheroidal structures growing in successive layers round a nucleus at rates of a few millimetres per million years, and reach average sizes of a few centimetres. They grow by precipitation both from overlying seawater and from pore waters in underlying sediment; and there is evidence that small organisms play some role in the nodule formation. Deep-sea nodules contain Co, Ni, and Cu in combined concentrations of up to 3%, which make them commercially attractive.

3 Manganese nodules also form in other environments, notably in continental margin regions. Here, they grow more quickly than their deep-sea counterparts, partly because manganese enriched in anoxic pore waters is more readily available for precipitation onto nodule surfaces. They contain lower concentrations of metals of commercial value.

4 Diagenesis encompasses reactions that occur between pore waters and solid phases, below the sediment surface, after the sediments become buried. The distinction between authigenesis and diagenesis is not always clear-cut. Diagenetic reactions result in the formation of new minerals, the liberation of some elements into pore waters, and the extraction of others from them; for example, magnesium into calcium carbonate, potassium into clay minerals, sodium into zeolites, and calcium into the pore waters.

5 There are considerable differences in the concentrations of redox-sensitive chemical species above and below the oxic–anoxic interface in deep-sea sediments. Below the interface, where conditions in the sediments are strongly anoxic, hydrocarbons form, and under favourable conditions of low temperature and high pressure, large accumulations of gas hydrates (clathrates) may develop.

Now try the following questions to consolidate your understanding of this Chapter.

QUESTION 5.7 What is likely to be the main source of manganese and iron in manganese nodules?

QUESTION 5.8 Which of the following statements are true, and which are false?

(a) The nepheloid layer is at its thickest where turbidity currents occur.

(b) Continental margin manganese nodules are a rich source of valuable metals.

(c) Throughout the oceans, conditions become anoxic at depths greater than about 1 m below the sediment surface.

(d) Both manganese and uranium are more soluble in reducing than in oxidizing conditions.

(e) Most material of aeolian origin escapes entrapment in deep-sea trenches.

APPENDIX THE GEOLOGICAL TIME-SCALE

Appendix The geological time-scale. Ages are given in millions of years (Ma).

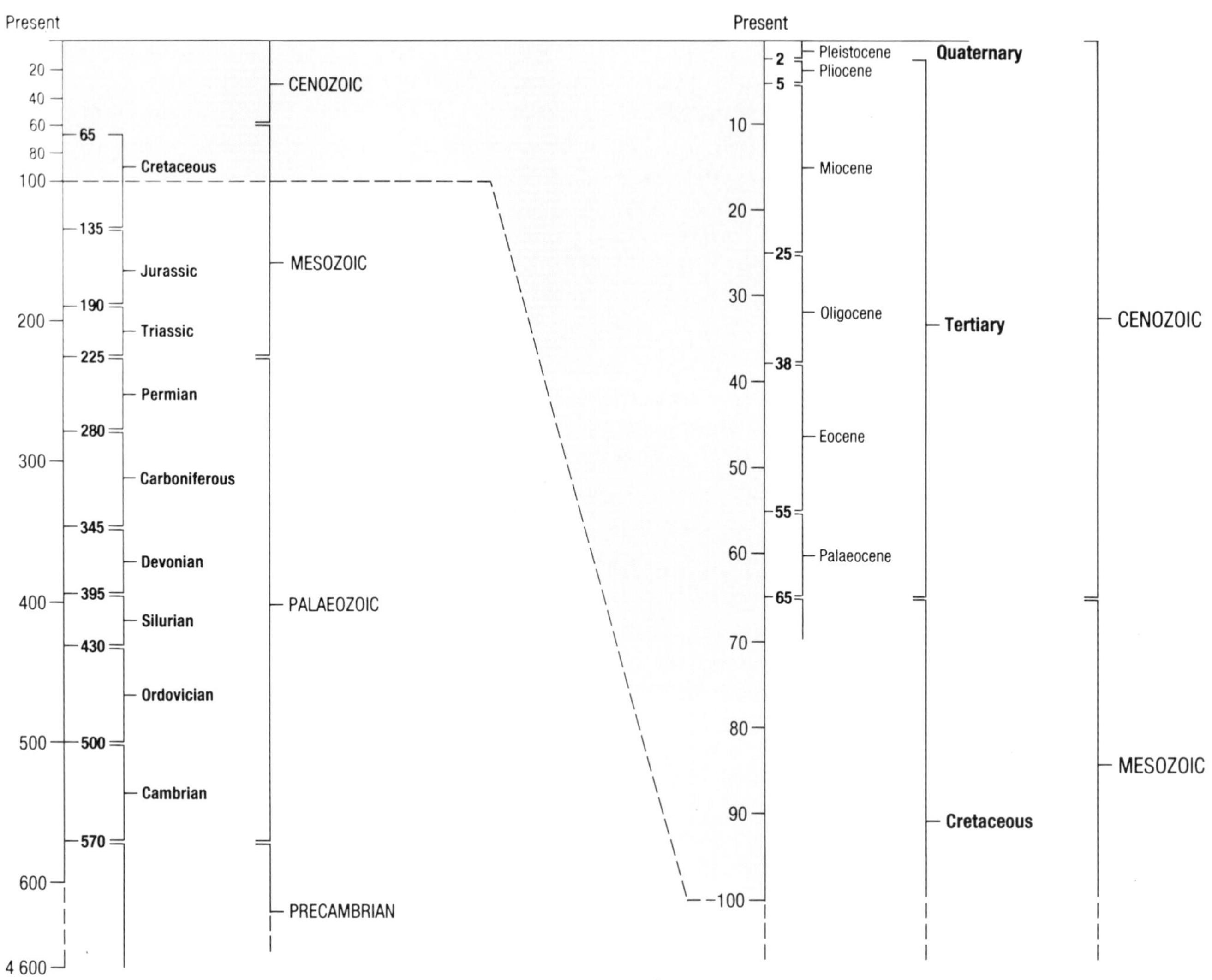

SUGGESTED FURTHER READING

BROECKER, W. S., AND PENG, T. S. (1983) *Tracers in the Sea*, Eldigio Publications. An advanced text describing the chemical and biological factors influencing the deposition and preservation of sediments.

HEEZEN, B. C., AND HOLLISTER, C. D. (1971) *The Face of the Deep*, Oxford University Press. A fascinating photographic record of animals, sediments and rocks on the sea-bed at all depths below the photic zone. The photographs are accompanied by a very readable explanatory text.

KENNETT, J. (1982) *Marine Geology*, Prentice-Hall. Substantial parts of this book are devoted to oceanic sediments and microfossils and to palaeoceanography.

MENZIES, R. J., GEORGE, R. Y., AND ROWE, G. T. (1973) *Abyssal Environment and Ecology of the World Oceans*, John Wiley. A biologically orientated book, containing much information about the interactions between deep-sea organisms and sediments.

MILLIMAN, J.C. (1974) *Marine Carbonates*. A comprehensive review of deep- and shallow-water carbonates, with a bias towards biological and geological aspects.

STOWE, K. (1983) *Ocean Science* (2nd edn), Wiley. A general text covering all aspects of oceanography, with emphasis on the multi- and interdisciplinary nature of this subject.

ANSWERS AND COMMENTS TO QUESTIONS

CHAPTER 1

Question 1.1 Calcareous biogenic sediments predominate along the ocean ridges which are in general the shallowest regions of the open oceans. Siliceous sediments are most abundant in the Southern Ocean, and occur also along the Equator in the Indian and Pacific Oceans, extending north and south along the eastern Pacific margin. The red clays are mostly in the deepest parts of the oceans, on the abyssal plains. (Mixed siliceous/red clay sediments occur in the Southern Ocean and in the North Pacific.)

Terrigenous sediments (gravels, sands, silts and clays) form the continental shelf–slope–rise sequences bordering the ocean basins (ice-rafted debris is carried by melting ice from polar regions).

These generalized distribution patterns will be discussed in more detail in later Chapters.

Question 1.2 (a) Diatoms are algae and depend upon light for photosynthesis. Light intensities are insufficient for net photosynthetic production (i.e. for algal growth) below the photic zone, which never extends deeper than about 200m, and is generally much shallower.

(b) In coastal waters and shallow seas, the photic zone is rarely more than 50m deep at best, so we would not expect to find benthic diatoms much below that.

Question 1.3 The surface living forms of both Foraminifera and Radiolaria are more spindly and delicate, and so are prone to greater dissolution than the deeper-water types. Their preservation potential is therefore less.

Question 1.4 Figure 1.14 shows a good correlation between contents of quartz and of illite in pelagic sediments from all the major ocean basins. Both result from continental weathering, both are found throughout the oceans, and the correlation suggests that neither is characteristic of weathering in a particular environment.

Question 1.5 As oceanic crust is formed at mid-oceanic ridges and moves away from them, its age increases with distance from the ridge. There is more time for sediments to accumulate, so thicknesses should be greater near continental margins than ridges. The second reason is that large quantities of sediment are deposited, direct from land, on the continental shelf and slope. Some of this sediment is carried down to the continental rise and the abyssal plains by turbidity currents.

Question 1.6 Yes. Diatoms and Radiolaria (Figures 1.11 and 1.12) deplete surface waters of silica to make their skeletons. Dissolution of some of the remains as they sink towards the sea-bed (and after they reach it) enriches the deep waters in silica. As you will see, the effect is much greater for Si than for Ca, because the oceanic reservoir of Si is so much smaller.

Question 1.7 (a) Kaolinite is produced by chemical weathering in low latitudes, so should be abundant in the equatorial Atlantic.

(b) Chlorite is characteristic of physical weathering in high latitudes, so should be abundant round Antarctica.

(c) Illite is a product of continental weathering in general, and there is much more land in the Northern than in the Southern Hemisphere.

CHAPTER 2

Question 2.1 (a) The profiles for nitrate and barium show that both these constituents are removed from solution in surface water and returned to solution in deeper waters ($NO_3^-:S$ and $Ba^{2+}:S$ ratios increase with depth). The straight profile for sodium suggests that there is no mechanism for removing the element from solution in the main body of the oceans. (The $Na^+:S$ ratio remains constant with depth); but see (c) below.

(b) Sodium is (i) conservative; nitrate and barium are (ii) non-conservative.

(c) Sodium is not *fixed* in organic tissue or skeletal material in the same way as phosphate and silica (and nitrate and barium) are. It is an essential ingredient of body fluids rather than part of the body fabric. We would die from lack or excess of salt: most marine organisms would perish in a lake, and most freshwater organisms would die in the sea simply because of the change in salinity. Sodium and other conservative constituents are by no means biologically inert, but their biological involvement relative to their concentrations in seawater is such that it does not show up in their concentration–depth profiles.

Question 2.2 (a) If the oceans take 500 years on average for a complete mixing or turnover cycle (i.e. for surface water to sink to the bottom and come back up again), then a dissolved constituent which spends only 100 years in solution will have been removed long before the cycle is complete.

(b) It must be non-conservative; its concentration in seawater cannot be determined by mixing processes alone—it is removed from solution long before mixing can be completed.

Question 2.3 The breakdown of marine snow is accomplished by decomposition (the dissolution of components and/or the biological 'glues'); disaggregation as a result of turbulence; and consumption by animals. The aggregates are lost from their place of formation by vertical settling and/or by horizontal advection in currents.

Question 2.4 (a) Well-stratified surface waters are gravitationally stable. Nutrients which sink from the surface cannot be replaced except by slow diffusion processes. If the water column is well mixed, on the other hand, nutrients sinking from the surface have a better chance of being carried back up again by turbulence.

(b) Silica is clearly being dissolved and recycled in upper parts of the water column (Figure 2.9 (c)). It forms skeletal material, however, which is more resistant to solution than soft organic tissue, and it therefore sinks deeper before significant solution occurs.

Question 2.5 (a) Table 2.1 gives a figure of 5.5×10^{-5} p.p.m. for the concentration of iron in seawater. The degree of enrichment in phytoplankton is thus considerable:

$$\frac{650}{5.5 \times 10^{-5}} = \frac{6.5 \times 10^{2}}{5.5 \times 10^{-5}} \approx 10^{7} \text{ times,}$$

and for zooplankton it will be about an order of magnitude less.

(b) For lead, data from Tables 2.1 and 2.2 give, for phytoplankton:

$$\frac{8}{2 \times 10^{-6}} \approx 4 \times 10^{6} \text{ times,}$$

and not much less for zooplankton.

These are dry weight enrichment factors, and would be about an order of magnitude less in the living organisms. Also, they represent averages, because different species concentrate different trace elements to different extents. Nonetheless, the concentration factors are very large.

Question 2.6 (a) Your sketch for inorganic germanium should resemble that for silica in Figure 2.9 (c), whereas for organic (methylated) germanium it should be a straight vertical line (like that for sodium in Figure 2.1).

(b) The profile for cadmium should resemble that for phosphate in Figure 2.9 (a).

You were not expected to put scales on your sketches, but for interest and information, Figure A1 (a) to (c) shows profiles for silica and germanium and the correlation between them, all for a single location. Figure A1 (d) shows a profile for cadmium.

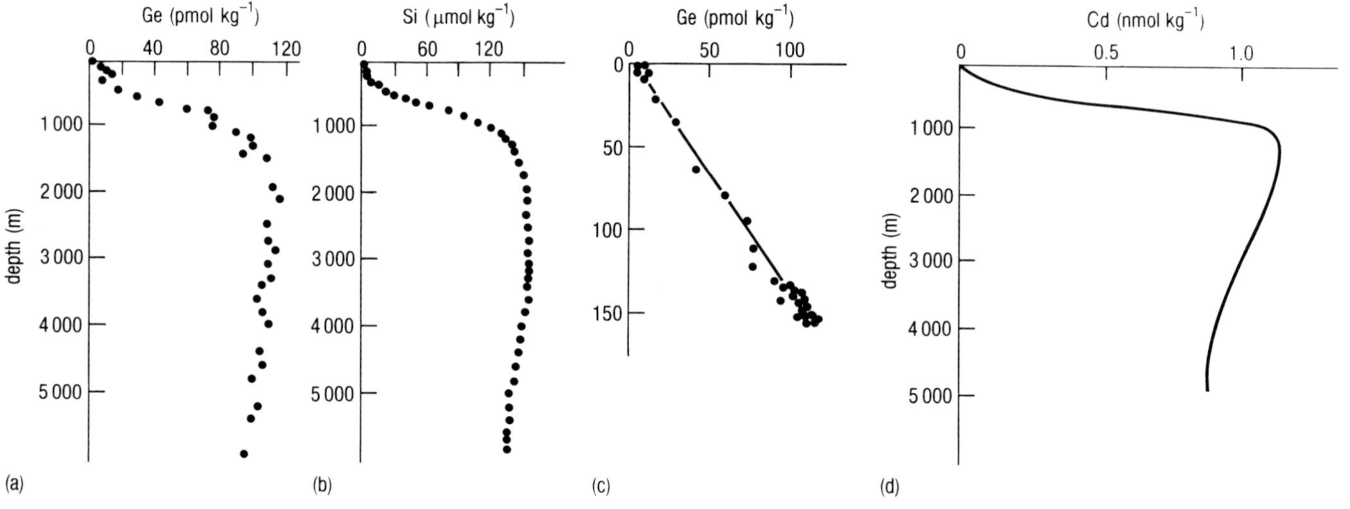

Figure A1 Concentration–depth profiles for (a) germanium (pmol = picomol = 10^{-12} mol) and (b) silicon at 20° N, 170°E in the north Pacific. (c) Correlation between Si and Ge in (a) and (b). (d) A concentration–depth profile for cadmium. (nmol = nanomol = 10^{-9} mol.) For answers to Question 2.6.

Question 2.7 (a) For profile (a), where the concentration decreases with depth from the surface and the location (off north-western Africa) is well away from the influence of major rivers, the source is probably mainly wind-blown dust from the Sahara. For profile (b), where the maximum concentration is at about 3.3 km depth, the only possible source is hydrothermal solutions from the mid-Atlantic ridge axis.

(b) The hydrothermal source is much the more important. The maximum concentration in profile (b) is more than 20 times greater than that in profile (a).

Question 2.8 For both (a) and (b) the elements are vertical neighbours in the Periodic Table. Strontium accompanies calcium in calcareous skeletal material (Section 2.1) and lies immediately beneath it in Group II of the Periodic Table. Germanium accompanies silicon in siliceous skeletal material (Figure A1) and lies immediately beneath it in Group IV. Note that uptake by analogy occurs only with *some* vertical neighbours. It does not apply to others, such as carbon and silicon.

Question 2.9 Assumption 2 ruled out the conservative constituents, which are not removed from seawater to any significant extent by biological activity. Assumptions 1 and 3 ruled out scavenged constituents, which have significant aeolian and hydrothermal inputs, and whose rates of input and removal vary considerably with time and from place to place, so that concentrations at any point change significantly with time.

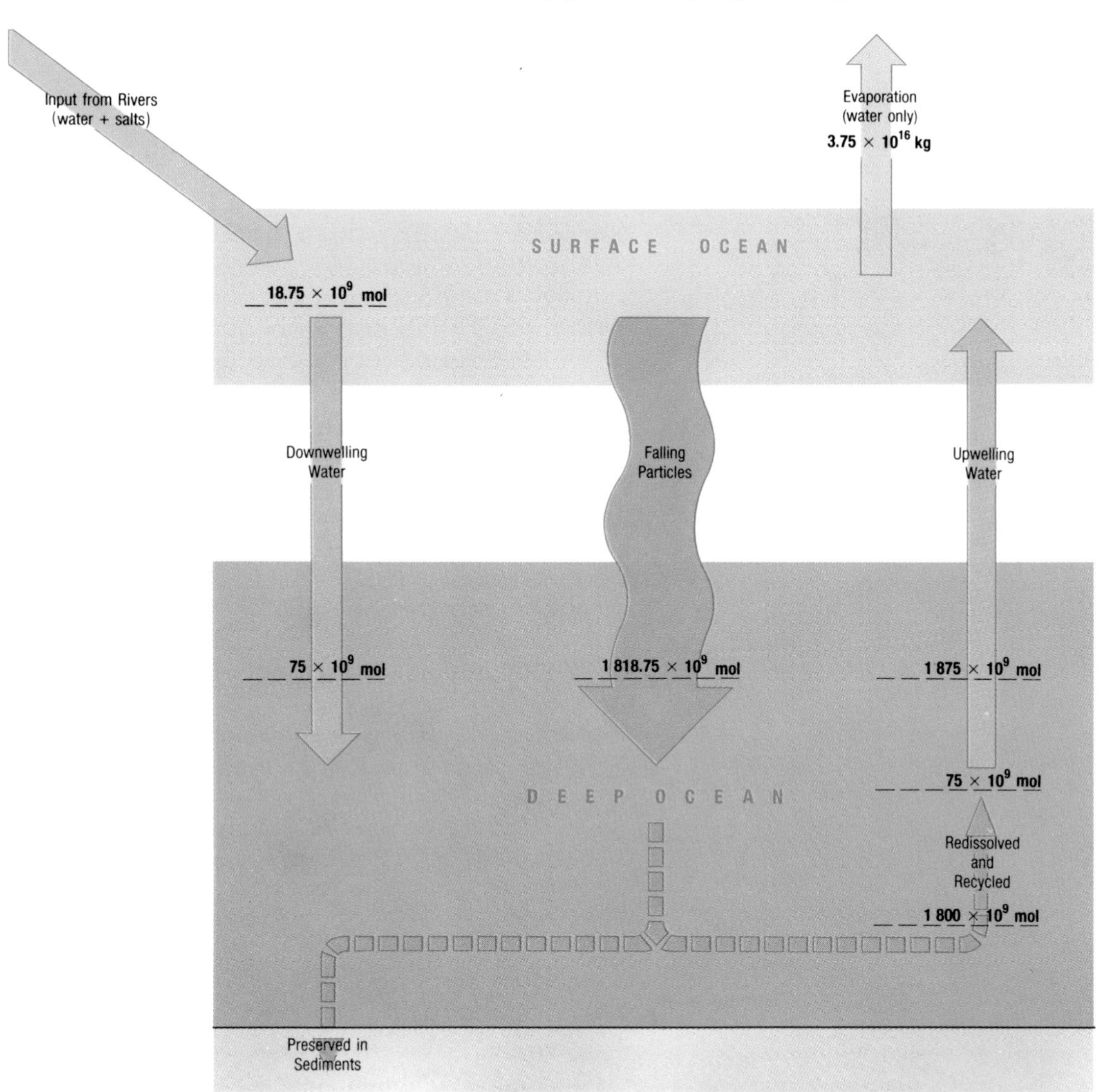

Figure A2 Completed Figure 2.16.

Question 2.10 For 1(a), we need to work out the following fraction as a percentage:

$$\frac{\text{Amount in Falling Particles}}{\text{Amount entering surface ocean (from Rivers and Upwelling)}}$$

$$= \frac{1818.75 \times 10^9}{(18.75 + 1875) \times 10^9} \times 100$$

$$= \frac{1818.75}{1893.75} \times 100 = 96\%$$

For 1(b), we need to work out the following fraction as a percentage:

$$\frac{\text{Amount Preserved in Sediments}}{\text{Amount entering surface ocean (from Rivers and Upwelling)}}$$

$$= \frac{18.75 \times 10^9}{1893.75 \times 10^9} \times 100 = 1\%$$

For 2, we need to work out the following fraction as a percentage:

$$\frac{\text{Amount Preserved in Sediments}}{\text{Amount in Falling Particles}}$$

$$= \frac{18.75 \times 10^9}{1818.75 \times 10^9} \times 100 = 1\%$$

Question 2.11 There is a striking difference. Nitrogen and phosphorus have the lowest ratios because they form soft organic tissue and are mainly recycled in surface waters (*cf.* Figure 2.3). Next comes the ratio for carbon, which forms both organic tissue and calcium carbonate; so more carbon reaches the deep ocean before it dissolves (bear in mind that much calcium carbonate escapes dissolution, however—Figure 1.4). The largest ratio is for silicon, which forms skeletal material that mostly dissolves in deep water (see Section 3.2).

Question 2.12 Profiles for scavenged elements show that concentrations *decrease* with depth. The older the water, therefore, the more time there is for elements in this group to be adsorbed and scavenged—the older the water, the lower the concentration becomes. For example, there is well over ten times more dissolved aluminium in the relatively young deep water of the Atlantic Ocean than in the older deep water of the Pacific.

Question 2.13 Argon is an inert gas and nitrogen is almost inert (*cf.* note 4 to Table 2.1), so both these gases can be considered to behave conservatively. Oxygen does not react with seawater, but it is used by organisms in respiration and so its concentration is changed by processes other than mixing; it is non-conservative. Carbon dioxide is also non-conservative, as it reacts with water, and so its concentration is also changed by processes other than mixing.

Question 2.14 (a) Values of w for (i) unreactive gases are generally less than unity, whereas for (ii) reactive gases they may be as high as several thousand. From the definition of w, therefore, rainfall will have much higher concentrations of reactive gases and much lower concentrations of unreactive gases than the surrounding atmosphere.

(b) Values of w can be as high as 10^5–10^6 for particles, but are only exceptionally as high as 10^5 for gases. This implies that in general rain scavenges particles more effectively than gases.

Question 2.15 The profile for pH follows that for oxygen quite closely, indicating that the water is most acid in the oxygen minimum layer. It is here that most of the oxygen has been used up in the formation of CO_2 (reverse of reaction 2.3, see the following text), which has combined with water molecules to form carbonic acid and its dissociation products.

Question 2.16 According to Figure 2.18, the oldest bottom waters are in the northern Indian and Pacific Oceans. As well as being enriched in nutrients we should expect them also to be most impoverished in dissolved oxygen.

Question 2.17 Profile (a) shows that the concentration of manganese is greater in the anoxic waters *below* the oxygen zero level than above. In profile (b), the manganese concentration is zero in the anoxic waters. Profile (a) therefore represents dissolved manganese (in the Mn(II) form), and profile (b) represents particulate manganese (Mn(IV)). Note that concentrations are greater for dissolved than for particulate manganese.

Question 2.18 There is more carbon in this ratio because for every mole of Ca in skeletal carbonate ($CaCO_3$) there is also a mole of C. The 25 moles of Ca must therefore be balanced by 25 'extra' moles of carbon, and $(105+25)=130$.

Question 2.19 The data for zinc show depletion in surface water, enrichment at depth, i.e. nutrient-type or recycled behaviour. Enrichment in deep water is greater at station II, so this is probably the Pacific station. We can confirm that with the cerium data, which show scavenged behaviour, i.e. depletion in deep water. There is less cerium in the samples from station II, which is consistent with it being in the Pacific (the older the water, the more time there is for particle-active elements to be adsorbed and scavenged).

Question 2.20 Average ocean-wide concentrations for scavenged elements cannot be very meaningful, because several have residence times *less* than the mean oceanic mixing time. Their distribution is uneven and reflects the influence of more or less localized sources.

Question 2.21 The two-box model can be applied to cyclic salts provided that the dissolved constituents concerned fall into the recycled category. If the Input from Rivers is corrected for the cyclic salt contribution, then the two-box model can provide an estimate of the proportion of 'new' material being removed to sediments each year.

Question 2.22 The concentration in surface waters can be used for ΔC in the equation because of the enormous difference between the actual concentrations of DMS and its equilibrium concentration, which is only a few ng $(10^{-9}\,g)\,l^{-1}$. There are enough uncertainties in gas flux calculations for the difference between, say, $300\,ng\,l^{-1}$ and $(300–3)\,ng\,l^{-1}$ to be neglected.

Question 2.23 (a) False. Most of them are, but carbon and, to a lesser extent, calcium, are two notable exceptions. Their involvement in biological processes is on a large-enough scale to affect their concentrations in seawater.

(b) False. These constituents are heavily involved in biological processes, and their concentrations are thus changed by processes other than mixing.

(c) False. Marine snow is abundant in some areas of high biological production, but particulate organic matter is generally dominated by faecal debris and bacteria.

(d) False. There cannot be more water coming up than is going down.

(e) False. Continental shelves are kept oxygenated by waves and tidal currents.

(f) True. In anoxic water, SO_4^{2-} is reduced to sulphide (reaction 2.5). There is less sulphate in solution, so the $SO_4^{2-}:S$ ratio must fall.

Question 2.24 The chemicals must have been emitted either as gases or as fine particulates. They were transferred across the air–sea interface by gaseous diffusion, by dry deposition or by washout in rain, as emissions from the stacks of the ships drifted across the nearby sea-surface. Once in the water column, the chemicals would be adsorbed and scavenged and transferred rapidly to the sea-bed, which lies at less than 200m depth over most of the North Sea. (As in the case of the Chernobyl nuclides (Section 2.2.3), such chemicals are more likely to accumulate to harmful levels in the sediments and affect the benthic species before animals inhabiting the water column are affected. The contamination may in fact be more widespread because no comparison sampling was done outside the area in question. In 1988, it was proposed that incineration at sea should cease by the mid-1990s.)

CHAPTER 3

Question 3.1 The ocean ridges are the shallowest parts of the open oceans, where temperatures are not so low and pressures not so high, and where calcite is less soluble.

Question 3.2 By substituting in equation 3.10:

For surface sample: $2.35-2.15=0.20\,\text{mol}\,\text{m}^{-3}$
For 4km deep sample: $2.45-2.40=0.05\,\text{mol}\,\text{m}^{-3}$

The surface sample plots well off the scale to the right, at the top of Figure 3.2, so the water is supersaturated with respect to calcite, which will not dissolve. The deep sample plots in the lower left of Figure 3.2, indicating that the water is undersaturated with respect to calcite, which should therefore dissolve.

Question 3.3 On all the profiles, the lysocline is the depth at which the profile begins to level out (at about 4km), i.e. where the amount of calcite in sea-floor sediments begins to decrease. The CCD is the depth at which the profile shows the calcite content to fall below 20%. The CCD on these profiles is the more variable, ranging from about 4.2 to 5km depth.

Question 3.4 The minimum in the $[CO_3^{2-}]$ profile at about 1 km depth coincides with the oxygen minimum layer, where oxygen consumption by respiration and bacterial decomposition is at a maximum (Figure 2.21). We can expect ΣCO_2 to reach a maximum also at this depth, and this in fact occurs in many regions.

Question 3.5 The present water depth at the site (Figure 1.7) is given as 1665 m, and the total sediment thickness is about 400 m. The top of the basaltic crust is therefore at a little over 2000 m depth. The site is thus well above the CCD at present, and the sediments in the core are all calcareous, so the site has been above the CCD since the Palaeocene. The presence of sandy and silty layers and wood fragments in the lower parts of the sequence suggest shallow-water conditions also. (For your information, glauconite is a mineral found only in shallow-water sediments; and volcanic ashes form in water depths of no more than 500 m.) The site must therefore have been quite shallow early in its history, and slowly subsided to its present depth. If the site is representative of the ridge as a whole, the inference is that the ridge has been above the CCD since it formed.

Question 3.6 The sea-bed is far below the depth at which the profile crosses the aragonite saturation curve (i.e. far below the aragonite lysocline). It is also some way below the calcite lysocline, and in fact at the sea-bed the value of $-\Delta CO_3^{2-}$ can be estimated to be about 15×10^{-6} mol l^{-1}, indicating substantial undersaturation and considerable dissolution of calcite. The sea-bed here could well be below the CCD (Figure 3.4 suggests that the CCD here is at about 4.5 km depth).

Question 3.7 Deep water is relatively cold and under pressure, and relatively rich in ΣCO_2 (e.g. Figure 3.1). The solubility of CO_2 gas in water decreases as (i) the pressure falls, and (ii) the temperature rises, in the upwelling water. Reaction 3.1 moves to the left, and the net flux of CO_2 is from sea to air (*cf*. Section 2.5.1).

CHAPTER 4

Question 4.1 The glacial sediments, because much of this material is likely to have been transported within ice and so protected from chemical attack.

Question 4.2 (a) By inserting the numbers into equation 4.1:

$$20 = 0.7\sqrt{\frac{(1.20-1.0)\times 10^3}{1.20\times 10^3}\times 9.8\times h_1}$$

$$= 0.7\sqrt{0.167\times 9.8\times h_1}$$

and, eliminating the square root by squaring both sides,

$$400 = (0.7)^2\times 1.64h_1$$

$$= 0.49\times 1.64h_1$$

$$= 0.8h_1$$

and re-arranging, we get

$$h_1 = \frac{400\,\text{m}}{0.8}$$

$$\approx 500\,\text{m}$$

(b) Neither. If (i) the density of the head were greater, then the $(\rho_t - \rho)/\rho_t$ term in equation 4.1 would be greater and the height would be less. Similarly, if (ii) the speed u_1 were lower, then as $u_1^2 \propto h_1$, the height would also be less.

Question 4.3 By inserting the numbers into equation 4.2:

$$20 = \sqrt{\frac{(1.20-1.0)\times 10^3}{1.20\times 10^3}\times 9.8\times 250\times \frac{\sin\beta}{6\times 10^{-3}}}$$

$$= \sqrt{\frac{9.8\times 250}{6\times 6\times 10^{-3}}\sin\beta}$$

and, eliminating the square root by squaring both sides,

$$400 = \frac{9.8\times 250}{36\times 10^{-3}}\sin\beta$$

Re-arranging:

$$\sin\beta = 400\times\frac{36\times 10^{-3}}{9.8\times 250}$$

$$= 5.9\times 10^{-3} = 0.0059$$

So $\beta = 0°\ 20'$ (or $0.33°$)

Question 4.4 The speed of the body is $0.5\,\mathrm{m\,s^{-1}}$, and the value of $\sin\beta$ is 10^{-5}. Substituting these numbers in equation 4.3 gives a value for the autosuspension limit,

$$v = \tfrac{2}{3}\, u_2 \sin\beta$$

$$= \tfrac{2}{3}\times 0.5\times 10^{-5}$$

$= 3.3\times 10^{-6}\,\mathrm{m\,s^{-1}}$ (compare *c.* $5\times 10^{-6}\,\mathrm{m\,s^{-1}}$ in preceding text)

Question 4.5 The statement is true for early stages of the flow, because the angle of slope controls the speed of the body (equation 4.2), which ‘feeds’ the head (Figure 4.5). It is false for later stages, where the angle of slope is less: the speed of the body falls, but that of the head is unaffected because it is primarily controlled by the height and the density contrast (equation 4.1).

Question 4.6 (a) Along Atlantic margins. There are deep ocean trenches around the Pacific, which trap sediment and prevent the formation of submarine fans.

(b) Increase. The more rapid the rise, the greater the rate of erosion. The more sediment delivered to the coast, the greater the number of turbidity currents.

Question 4.7 False. Most abyssal plain sediments are deposited from turbidity currents (*cf.* Figure 4.6 (d)). Hence their absence from most of the Pacific (Figure 4.9).

CHAPTER 5

Question 5.1 Suspended sediment concentrations are greatest along the western Atlantic margin. In the southern Atlantic, the northward flow of Antarctic Bottom Water is strongest along the western boundary. The southward flow of North Atlantic Deep Water along the western side of the northern Atlantic is almost as strong.

Question 5.2 There is a broad correlation between rich manganese nodule fields and regions of the sea-floor where the sediments are low in organic carbon, which is consistent with the presence of well-oxygenated bottom waters. Figure 2.8 suggests that nodules should also occur in the Indian Ocean, and they do. The nodule field in Figure 5.8 *is* in the Indian Ocean.

Question 5.3 The concentrations of nickel, copper and cobalt are both individually and collectively higher in nodules from the Pacific than from the other two oceans, according to Table 5.1. That would seem to suggest that the Pacific should be the main target for exploration, especially as most nodule fields are there. However, Table 5.1 contains averages only, which may conceal wide variations, and there may be rich nodule fields in the other oceans also.

Question 5.4 Nodules from continental margins differ from other marine manganese nodules in being very rich in manganese and poor in iron (high Mn:Fe ratios) and in having very low concentrations of other metals, especially Co.

Question 5.5 (a) The profiles show that both Mg^{2+} and K^+ are depleted in the pore waters of this sequence, whereas Ca^{2+} and SiO_2 are enriched, as depth increases; although for Mg^{2+}, Ca^{2+} and SiO_2 there is little change below about 250m.

(b) Dissolved silica is greatly enriched in the pore waters below about 200–250m, where the sediments contain a higher proportion of biogenic siliceous material.

Question 5.6 (a) The concentration–depth profiles for Al_2O_3 and SiO_2 (the principal chemical components of clay minerals) in the sediment core are virtually straight vertical lines. If there were significant variations with depth in the content and type of clay minerals in this core, then these profiles would not be straight lines.

(b) The profile shows that dissolved oxygen in the pore waters decreases to virtually zero at the transition. The profile for nitrate has the same form, because in the absence of oxygen it is also used as an oxidizing agent in the same way that sulphate is (reaction 2.5).

(c) Both these redox-sensitive species are soluble in anoxic (reduced) waters. They go into solution below the transition and their concentration therefore rises with depth.

Question 5.7 Hydrothermal activity is probably the main source, as it supplies more iron and manganese to the sea than rivers do.

Question 5.8 (a) False, in general. Turbidity currents may locally intensify the nepheloid layer, but increased eddy kinetic energy in both surface and deep currents is the major cause of the nepheloid layer.

(b) False. They grow too fast to accommodate these metals.

(c) False. The example illustrated in Figure 5.16 is oxygenated throughout.

(d) False. Manganese *is* more soluble, but uranium is *less* soluble in reducing conditions.

(e) True. Most aeolian material is carried far out into the ocean basins, away from continental margins.

ACKNOWLEDGEMENTS

The Course Team wishes to thank the following: Dr. Martin Angel and Dr. Dennis Burton (the external assessors); Dr. Mike Whitfield for helpful discussion and advice on Chapter 2; and Mr. Mike Hosken and Mrs. Mary Llewellyn for advice and comment on the whole Volume. Dr. Ric Jordan provided several of the electron micrographs in Chapter 1.

The structure and content of the Series as a whole owes much to our experience of producing and presenting the first Open University course in Oceanography (S334), from 1976 to 1987. We are grateful to those people who prepared and maintained that Course, to the tutors and students who provided valuable feedback and advice and to Unesco for supporting its use overseas.

Grateful acknowledgement is also made to the following for material used in this Volume:

Figures 1.1, 1.2 and 5.7 Institute of Oceanographic Sciences; *Figure 1.4* R.N. Anderson (1986) *Marine Geology*, John Wiley and Sons; *Figure 1.4 and base of Figure 4.9* NASA; *Figure 1.6* Paul Yates, Open University; *Figure 1.7* B.P. Luyendyk and T.P. Davies (1974) in *Initial Reports of the Deep Sea Drilling Project*, **26**, US Government Printing Office; *Figure 1.8(a)* A. McIntyre, Lamont-Doherty Geological Observatory; *Figures 1.8 (b)–(d), 1.9(b), 1.11(b) and 1.12(b)* R.W. Jordan and W. Smithers, University of Surrey; *Figures 1.9(a) and 1.12(a)* D. Breger, Lamont-Doherty; *Figure 1.10* J.D. Milliman (1974) *Recent Sedimentary Carbonates*, Part 1, Springer-Verlag; *Figure 1.11(a)* L. Burckle, Lamont-Doherty; *Figure 1.13* W.G. Deuser, Woods Hole Oceanographic Institution; *Figure 1.14* H.L. Windom, in J.P. Riley and R. Chester (eds.)(1976) *Chemical Oceanography*, Vol. 5, Academic Press; *Figures 2.2(b)–(c), 5.1, 5.2 and 5.14* Bruce Heezen; *Figures 2.4 and 2.5* A.L. Alldredge, University of California, Santa Barbara; *Figure 2.6* A.L. Alldredge and M.W. Silver (1988) in *Progress in Oceanography*, **20**, Pergamon; *Figure 2.7* R. Lampitt, IOS Deacon Laboratories; *Figure 2.10* E. Suess (1988) in *Nature*, **333**, Macmillan; *Figure 2.11* K.J. Orians and K.W. Bruland (1986) in *Earth and Planetary Science Letters*, **78**, Elsevier; *Figure 2.12(b)* G. Klinkhammer *et al.* (1986) in *Earth and Planetary Science Letters*, **80**, Elsevier; *Figure 2.14* W.S. Moore and J. Dymond (1988) in *Nature*, **331**, Macmillan; *Figures 2.15, 2.17–2.19 and 3.1* W.S. Broecker (1974) *Chemical Oceanography*, Harcourt Brace Jovanovich Inc.; *Figure 2.21* H. Friedrich (1969) *Marine Biology*, Sidgwick and Jackson; *Figure 2.22* J.P. Riley and R. Chester (1971) *Introduction to Marine Chemistry*, Academic Press; *Figure 2.24* S.E. Calvert and N.B. Price (1970) in *Nature*, **227**, Macmillan; *Figures 3.2, 3.3 and 3.6* W.S. Broecker and T.S. Peng (1982) *Tracers in the Sea*, Lamont-Doherty; *Figure 3.4* A.P. Lisitzin (1972) *Spec. Pub. Soc. Econ. Paleon. & Miner.*, Tulsa, **17**; *Figure 3.7* J.G. Sclater and D.P. McKenzie (1973) *Geological Society of America Bulletin*, **84** (10), Geological Society of America; *Figure 3.8* J.W. Murray (1988) in *Journal of the Geological Society*, **145** (1), Blackwell; *Figure 4.2* R.W. Embley (1976) in *Geology*, **4**, University of Chicago Press; *Figure 4.3(a)* D.B. Prior, Louisiana State University; *Figure 4.3(b)* D.K. Nagel, San José State University, H.T. Mullins, Syracuse University, and H.G. Green, USGS: *Figure 4.4(a)* H.W.

Menard (1964) *Marine Geology of the Pacific*, McGraw-Hill; *Figure 4.4(b)* K.K. Turekian (1976) *Oceans,* 2nd edn, Prentice-Hall; *Figure 4.5(a)* J. Best, University of Hull; *Figure 4.6(d)* P.P.E. Weaver and R.G. Rothwell; *Figure 4.7* I.N. McCave and K.P.N. Jones (1988) in *Nature*, **333**, Macmillan; *Figure 4.8* B.G. Heezen and C.D. Hollister (1971) *The Face of the Deep*, Oxford University Press; *Figure 5.3* P.E. Biscaye and S.L. Eittreim (1977) in *Marine Geology*, **23**, Elsevier; *Figure 5.4* C.D. Hollister and I.N. McCave (1984) in *Nature*, **309**, Macmillan; *Figure 5.5* T.F. Gross and A.J. Williams III, Woods Hole Oceanographic Institution; *Figures 5.6 and 5.17* L.B. Sand and F.A. Mumpton (eds.)(1978) *Natural Zeolites*, Pergamon; *Figure 5.10* Courtesy of R.K. Sorem and A.R. Foster, Washington State University; *Figures 5.11–5.13* J. Greenslate, Scripps Institution of Oceanography; *Figure 5.15* B.J. Skinner and K.K. Turekian (1973) *Man and the Oceans*, Prentice-Hall; *Figure 5.16* E.A. Perry *et al.* (1976) in *Geochimica and Cosmochimica Acta*, **40**, Pergamon; *Figures 5.18 and 5.19* J. Thomson *et al.* in P.P.E. Weaver and J. Thomson (eds.)(1987) *Geology and Geochemistry of Abyssal Plains*, Geological Society; *Figure A1* P.N. Froelich Jr. and M.O. Andreae (1981) in *Science*, **213**, American Association for the Advancement of Science.

INDEX

Note: page numbers in italics refer to illustrations; in bold to tables

THE OCEANOGRAPHY COURSE TEAM

Authors
Joan Brown
Angela Colling
Dave Park
John Phillips
Dave Rothery
John Wright

Editor
Gerry Bearman

Design and Illustration
Sue Dobson
Ray Munns
Ros Porter
Jane Sheppard

This Volume forms part of an Open University course. For general availability of all the Volumes in the Oceanography Series, please contact your regular supplier, or in case of difficulty the appropriate Pergamon office.

Further information on Open University courses may be obtained from: The Admissions Office, The Open University, P.O. Box 48, Walton Hall, Milton Keynes, MK7 6AA.

Cover illustration: Satellite photograph showing distribution of phytoplankton pigments in the North Atlantic off the US coast in the region of the Gulf Stream and the Labrador Current. *(NASA and O. Brown and R. Evans, University of Miami.)*

OCEAN CHEMISTRY AND DEEP-SEA SEDIMENTS